Renewable Energy in Power Systems

Renewable Energy in Power Systems

David Infield
Department of Electronic and Electrical Engineering
University of Strathclyde
Glasgow, UK

Leon Freris
Centre for Renewable Energy Systems Technology (CREST)
Loughborough University
Leicestershire, UK

Second Edition

Registered Offices
John Wiley & Sons, Inc., 111 River Street, Hoboken, NJ 07030, USA
John Wiley & Sons Ltd, The Atrium, Southern Gate, Chichester, West Sussex, PO19 8SQ, UK

Editorial Office
The Atrium, Southern Gate, Chichester, West Sussex, PO19 8SQ, UK

For details of our global editorial offices, customer services, and more information about Wiley products visit us at www.wiley.com.

Wiley also publishes its books in a variety of electronic formats and by print-on-demand. Some content that appears in standard print versions of this book may not be available in other formats.

Library of Congress Cataloging-in-Publication Data
Names: Infield, D. G., author. | Freris, L. L., author.
Title: Renewable energy in power systems / David Infield, Leon Freris.
Description: Second edition. | Hoboken : Wiley, 2020. | Includes
    bibliographical references and index.
Identifiers: LCCN 2019030493 (print) | LCCN 2019030494 (ebook) | ISBN
    9781118649930 (cloth) | ISBN 9781118788585 (adobe pdf) | ISBN
    9781118788561 (epub)
Subjects: LCSH: Renewable energy sources.
Classification: LCC TJ808 .F76 2019 (print) | LCC TJ808 (ebook) | DDC
    621.042–dc23
LC record available at https://lccn.loc.gov/2019030493
LC ebook record available at https://lccn.loc.gov/2019030494

Cover Design: Wiley
Cover Image: © photography by Spencer Bowman/Getty Images

Set in 10/12pt WarnockPro by SPi Global, Chennai, India
10 9 8 7 6 5 4 3 2 1

Contents

Foreword

Congratulations! By holding this book (or e-read equivalent) in your hands, you are very likely showing an interest in the engineering of renewable energy concepts. (And well done, too, on reading a Foreword, most people, including myself, jump straight into the technical chapters but that's another story.)

I want to urge you to keep that interest in renewable energy going. When I studied electrical engineering in the 1980s, anything renewable (wind, solar, and to some degree hydro) was classified by many scholars, industry practitioners and stakeholders as Alternative Energy. 'Alternative' to the might of having a standardised high voltage and low voltage grid network that delivered electricity to homes and factories from bulk sources such as coal, oil, gas and nuclear power stations. We had some hydro but this was pretty much exclusively used for peak demand management and frequency control, and virtually no other renewable energy sources existed at scale. That 'Alternative' badge pretty much also made anyone who was trying to explore renewable sources be tagged as anything from 'non-mainstream' to 'tree hugging' to 'just plain crazy'. It was difficult to be different and to do anything other than the decades old ways of generating and delivering electricity. To be fair, this was for good reason, of course – when the paying of your energy bills formed a significant part of just about everyone's income it was often the cheapest way of generating electricity and delivering it reliably to 'consumers'. Implementing significant change was deemed difficult and costly. It was easier to move in straight lines.

Now, we are in a very different age. I'm pleased the term 'consumer' has almost gone, and has been replaced with the word 'customer'. The production and sale of electricity is now much more competitive and customer value is replacing monopoly product and service. Regrettably, the world still has millions of people struggling to afford to pay for energy and so the changes we want to make to improve our environment must continue to take affordability into account. Often, this requires innovative ways of doing things, finding new solutions to use existing infrastructure and challenging the age-old engineering and regulatory norms that dominate power systems around the world.

Further, the steady realisation that those who warned about the effects of noxious gases that created acidic rain, and then CO_2 contributing to climate change, were actually right as opposed to crazy, has led to global legislation to do something about it. Renewable energy, particularly wind and solar power, has made great strides. The Alternative Energy world is on a journey to becoming far more mainstream. There will always be some who deny climate change, of course. Nevertheless, what is indisputable is that fossil fuels are finite, and it's better, in my view, to generate power using sustainable resources and to either leave our fossil fuels alone or at least find far more efficient, non-polluting uses for them.

In my undergraduate days, we were taught only to be engineers. We entered the world of work with maths and physics bursting out of our heads – only to find that most places of work required, in addition to one's engineering talents, a head for economics, environmental awareness, customer service and stakeholder management. These are qualities I'd urge all engineers to be aware of if you want to see your work make the best progress.

The rapid development of renewable energy sources that are, by their nature, distributed to take advantage of local energy resources rather than being centrally located for other economic reasons, makes the future creation and delivery of electricity (and other energy products) far more challenging to forward plan but it also makes it much more interesting and exciting for tomorrow's engineers. We no longer move forwards in straight lines of "bigger equals better". We have to think tangentially, innovatively and with a healthy mixture of focus on detail where it's due but also on wider system and regulatory imperatives. The age old rules of 'this is how it's done' need not always apply.

I'm delighted to see the growth of wind power, especially offshore, move forwards so rapidly in recent years, and installed volumes are forecast to grow heavily over the next couple of decades. Wind 'power stations' are now at a competing scale with fossil fuel plants, costs are competitive and continue to fall and reliability is ever improving. Yet, there is more to be done to keep those improvements going, using new materials to improve sustainability, improving designs to make turbines lighter, blades more aerodynamic, foundations lighter yet stronger, and cables more sea reliant. Cost-effective deep-sea solutions such as floating wind are now within our grasp. The digital age is creating more prospects for operations and maintenance to be done remotely and with less human intervention. Fewer people working in situ means a safer industry and human power will be used more to control, to innovate and to invent. Tidal, wave and even kite power solutions continue to follow on as potential sources for the future, alongside many concepts in early stages of development such as carbon from air extraction.

We need to think, too, of how our grid systems and its regulations, designed decades ago for very different ways of working, will evolve at the same time, likely with more smart grid technology, greater interconnectivity both onshore and to offshore plants as well as to other countries, with greater storage of likely not one but a few types (batteries, pumped storage, mechanical and maybe compressed air, for example). Wind and other forms of renewables are obvious candidates for providing renewable energy for heating and for transportation, finally providing a first class opportunity to decarbonise those latter sectors which, in fact, are much larger users of energy than that of electricity in our homes.

There are prospects for using renewables to process sea water into hydrogen or ammonia – creating sustainable, transportable, storable fuels. These products could provide sensible solutions to the continued usage of our electricity and gas grid networks, not to mention the highway refuelling of vehicles. When we create new renewable electricity solutions in the future we should do so with heat and transport in mind, all such systems require major innovations to decarbonise.

Finally, one other big change is clearly happening in our world of renewable energy – that of people power. The voice of the customer, especially through social media, is now a far more powerful tool in policy making and in popularising new technological successes (the latest electric cars, for example) – as well as reputation breaking when things go wrong. We must bring change to the ways we generate and use our energy sources, but we must always have customers in mind. They are key drivers to better cost reduction, energy usage efficiency and sustainability.

So, what does all this mean? It means engineering challenge and plenty of innovation opportunity. That's a great thing for new engineers. This book, written by two of the most gifted professors who are as passionate about the world of renewable energy as I have ever had the pleasure of meeting, will provide a solid foundation for you to move forwards. It's up to you as to whether you move forwards in straight engineering lines or in more innovative ways!

Andrew Jamieson
Chief Executive, ORE Catapult

Preface to the First Edition

There is worldwide agreement on the need to reduce greenhouse gas emissions, and different policies are evolving both internationally and locally to achieve this. On 10 January 2007, the EU Commission announced an Energy Package that was endorsed by the European Council. The objectives are that, by 2020, EU greenhouse gases are to be reduced by 30% if a global agreement is arrived at or by 20% unilaterally. One of the vital components in the achievement of this goal is the intention to provide a 20% share of energy from renewable energy (RE) sources in the overall EU energy mix.

At present, wind power is the leading source of new renewable energy. World wind power capacity has been growing rapidly at an average cumulative rate of 30% over the last 10 years. About 20 GW of new capacity was installed in 2007, bringing the world total in that year to 94 GW. This annual investment represents around €25 billion by an industry that employs 200, 000 people and supplies the electricity needs of 25 million households. This considerable expansion has attracted investment from major manufacturing companies such as General Electric, Siemens, ABB and Shell, as well as numerous electricity utilities, notably, E.ON and Scottish Power. The future of wind power over the next two decades is bright, indeed.

Generation of electricity from the sun can be achieved directly using photovoltaic (PV) cells or through solar concentration to raise steam and drive conventional turbines. Over the last few years, considerable progress has been made in the reduction of the cost of PV generated electricity, with 2006 seeing the total value of installed capacity reaching €15 billion and with cell global production in that year approaching 2.5 GW. It is expected that further technology improvement and production cost reduction over the next decade will result in wide scale competitive generation from this source.

Marine energy is an exciting, but less well-developed technology. Tidal barrages, tidal stream turbines and wave energy devices are all in the experimental and pre-commercial stage but are expected to make a significant contribution by around 2015. Geothermal energy is now established in countries like Iceland, with a significant accessible resource, and as the technology develops could be taken up more widely. Last but not least, there are bioenergy and biofuels, important because they offer many of the advantages of fossil fuels, in particular, being easily stored. Not surprisingly, they are receiving much attention from policy makers and researchers both in the EU and North America.

Most of this renewable energy will be converted into electricity. The renewable energy resource will be geographically highly distributed and, being mostly dependent on changing weather and climate, cannot be directly controlled in the way fossil fuelled generation is. Electrical power networks were designed to operate from electricity generated in a few large power stations fuelled by coal, gas or uranium fuels, readily available on the international market and, to varying degrees, controllable. Significantly increasing the input from renewable energy sources requires a revision of the way power systems are designed and operated in order to accommodate these variable sources better. This book is an introduction to this important topic.

The material in this book is largely based on a Master's course module taught for over 10 years at the Centre for Renewable Energy Systems Technology (CREST) at Loughborough University. The course, as a whole, was designed to provide general technical education in all major electricity generating renewable energy sources and their integration in electrical networks. Students taking this course normally have first degrees in numerate topics ranging from Physics or Engineering to Environmental Science. The course modules are, therefore, designed for students who, although they may be very knowledgeable in their speciality, will only have elementary knowledge of other topics.

Likewise, this book assumes no previous knowledge in power systems engineering and guides the reader through the basic understanding of how a power system is put together and the way in which it ensures that the consumer demand is met from instant to instant. The characteristics of traditional and renewable energy (RE) resources are described with special reference to the variability of the latter and the way this impacts on their utility. These resources are available in a form that either has to be converted into electricity and/or their electrical output has to be conditioned before it can be fed into the grid. The book covers these aspects and stresses the importance of power electronic technology in the process of power conditioning. The power flows in an electricity network have to be appropriately controlled and the book addresses the way this is achieved when these new sources are integrated. The economics of renewable sources will determine their take-up by the market, and this issue is also addressed, and in some detail. Finally, an eye is cast on the future development of RE technologies and the way that power systems may evolve to accommodate them. An Appendix is available for readers who require a more mathematical coverage of the way electricity is generated, transported and distributed to consumers.

Preface to the Second Edition

Since the first edition of this book was published in 2008, renewable energy technology and its deployment has advanced in a spectacular manner. Much of this development was actively supported by national and international policies. Of particular note was the framework provided by the European Union back in 2007 to target a 20% contribution from renewables by 2020 across the EU member states. In 2008, the fastest growing renewable energy source was wind power, but today a comparable and impressive growth is being demonstrated by photovoltaics. Even in the UK, which has a limited solar resource, these two technologies together have almost squeezed coal fired power generation to extinction.

However, as renewable power generation has become more widespread, so has its impact on power system operation. In that sense, this book has become more important and we hope that it will be useful to students and researchers in helping them to understand the key issues that the power industry will need to deal with in the years ahead.

As with the first edition, the characteristics of the renewable energy sources are described with stress on those factors that impact on their integration into power systems. New material has been included that explores the potential for renewable energy sources to contribute to power system dynamic regulation through the provision of system frequency response; such capability is now commercially available, and is critical for power system stability as the renewable energy penetration increases.

The content of this book is still largely based on a Master's course module taught at the Centre for Renewable Energy Systems Technology (CREST) at Loughborough University, when both authors worked there. The course, as a whole, was designed to provide a general technical introduction to the integration of renewable energy generation sources into power systems. The Master's students at Loughborough had normally obtained degrees in technical and numerate subjects but not necessarily in electrical engineering. The book, likewise, assumes no detailed knowledge of the topic and an Appendix covering the basics of electrical power engineering is included to support readers with no background in this area. The number of worked examples has been increased to help readers improve their quantitative capabilities.

The final chapter, covering future developments, has been expanded to cover technology advancements over the last decade. It draws on research from around the world to indicate where the different renewable technologies are heading, and the potential for them to support a totally CO_2 free and sustainable electricity system in the future.

March 2019   *David Infield and Leon Freris*

Acknowledgements

This second edition includes much material from the first that was developed by colleagues from CREST at Loughborough University when both the authors were working there. In particular, Dr Murray Thomson provided much of the power electronics material of Chapter 4 and most of the contents of Chapters 5 and 6. His comments and criticisms during the initial development of the book were invaluable and strongly informed the structure of the first edition, which has been retained. Professor Simon Watson of CREST, now with Delft University of Technology, provided key contributions to Chapter 7. In addition, the dynamic demand control example in Chapter 3 was taken from a CREST Master's dissertation by J. A. Short. The illustration of frequency response contributions from wind was taken from the PhD work of Lei Wu while at the University of Strathclyde, now a Lecturer at Nanjing Institute of Technology. The many other researchers and organisations who allowed us to reproduce material are acknowledged in the text.

We are also grateful to the Wiley staff at Chichester and elsewhere who have supported us in the production of the manuscript.

Last, but certainly not least, we would like to dedicate this edition, as we did with the first edition, to our respective partners, Marion Peach and Delphine Freris, who have accepted the times we were often distracted from family life by the demands of the writing.

About the Companion Website

This book is accompanied by a companion website:

www.wiley.com/go/infield-renewable-energy

The website includes:
   Document with updates on renewable energy.

   Scan this QR code to visit the companion website.

1

Energy and Electricity

1.1 The World Energy Scene

1.1.1 History

Energy has played a key role in the development of society. In the pre-industrial world, this energy was provided mostly by man and animal power and from the burning of wood for heating, cooking and smelting of metals. Coal, through the then new technology of steam engines, mechanised production, improved transportation and powered the industrial revolution. The inter-war years saw the rise of oil exploration and use. Access to this critical fuel became a key issue during the Second World War. Post-war industrial expansion and prosperity was increasingly driven by oil, as was the massive growth in private car use. More recently, a new phase of economic growth has been underpinned by natural gas; most recently much of this derived by fracking. Nuclear power makes a contribution to electricity generation. However, transport still depends almost entirely on oil.

A substantial proportion of coal and gas production, and all nuclear power, is used to generate electricity, which has been widely available now for over a century. Electricity is a premium form of energy due to its flexibility and ease of distribution and use. Demand worldwide continues to grow driven by increased industrial activity and the widening of access to electricity from consumers in the developing world.

1.1.2 World Energy Consumption

The present global yearly *primary energy*[1] consumption is in round figures about 550 EJ[2] This is equivalent to about 1.5×10^{17} Wh or 150 000 TWh. Dividing this figure by the number of hours in the year gives 17 TW or 17 000 GW as the average rate of world primary power consumption. The pie chart in Figure 1.1 shows the percentage contribution to world primary energy from the different energy sources according to data taken from the International Energy Agency (IEA) Key World Energy Statistics, 2016 [1].

---

1 Primary energy is the gross energy before its transformation into other more useful forms like electricity.
2 The unit of energy in the MKS system is the joule, denoted by J. Multiples of joule are kJ, MJ, GJ, TJ (T for tera denoting 10^{12}) and EJ (E for exa denoting 10^{18}); the unit of power is the watt (W) and represents the rate of work in joules per second. Electrical energy is usually charged in watt-hours (Wh) or kWh. Joules can be converted into Wh through division by 3600. Energy resource and use is also often presented in million tons of oil equivalent (Mtoe) where I TWh = 0.086 Mtoe.

Renewable Energy in Power Systems, Second Edition. David Infield and Leon Freris.
© 2020 John Wiley & Sons Ltd. Published 2020 by John Wiley & Sons Ltd.
Companion website: www.wiley.com/go/infield-renewable-energy

Biofuels and waste
10.3%

Hydro 2.4%

Nuclear 4.8%

Natural gas
21.2%

Oil
31.3%

Coal[2]
28.6%

Other[3] 1.4%

Figure 1.1 Percentage contribution to world primary energy in 2014 ([2]peat and oil shale included with coal,[3] other includes the remaining renewable energy sources, wind, solar, geothermal and tidal).

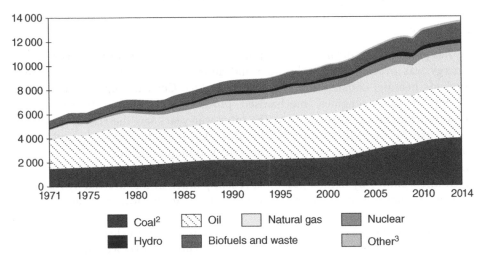

Coal[2]   Oil   Natural gas   Nuclear

Hydro   Biofuels and waste   Other[3]

Figure 1.2 Recent growth in world primary energy supply by fuel in Mtoe ([2]peat and oil shale included with coal,[3] other includes the remaining renewable energy sources, wind, solar, geothermal and tidal).

World energy supply continues to increase steadily as can be seen in Figure 1.2 (also taken from the IEA Statistics 2016), with most of this being accounted for by an increase in fossil fuels, particularly coal and gas. The rapid development of the economies of India and China is contributing significantly to this growth, despite currently much lower per capita energy consumption in these countries than in Europe and North America.

1.1.3 Finite Resources

It is extremely difficult to determine precise figures on the ultimate availability of fossil fuels. According to the major oil and gas companies, significant new resources of oil are still being developed or remain to be discovered. A safe assessment is that there is enough oil from traditional sources to cover the present demand for at least 30 years. The latest figures for global gas reserves indicate that these are approximately 50% higher than for oil, at some 60 years of current demand. And gas is far less explored than oil so there is probably more to be found. There are, however, unconventional hydrocarbon resources – such as heavy oil and bitumen, oil shale, shale gas and coal bed methane – whose total global reserves have been assessed very roughly to be three times the size of conventional oil and gas resources. These are more expensive to extract but are now being exploited as the price of fossil fuels increases due to the steady depletion of the more easily accessible reserves. Most notable has been the recent increase in shale gas extraction in the USA derived through the process known as fracking. Fortunately for fossil-fuel dependent economies, coal reserves are considered to be many times those of oil and gas and could last for hundreds of years. The down-side of coal is its high carbon content, a topic to be discussed later.

Much debate is currently focused on when the so called, peak oil and gas might occur. This is when the oil and gas extraction rate starts to fall, and occurs well before resources run out. It is important because it signals that demand will most likely not be fully met with prices rising significantly as a consequence. Certainly, the UK's North Sea reserves of oil and gas are fast declining with peak extraction having already occurred in 2003. Given the enormous investment in extraction and supply infrastructure, and the profits to be made, it would be surprising if those with vested interests did not work hard to maintain confidence in these sources.

Fuel for nuclear fission is not unlimited and several decades ago this prompted interest in the fast breeder reactor that, in effect, extends the life of the fuel. However, the political dangers inherent in the fast breeder cycle, with its production of weapons grade plutonium, has limited its development to a few prototype reactors that had major operational problems and are now decommissioned. The lifetime of uranium reserves for conventional fission at current usage has been estimated by some as around 50 years but such calculations are very dependent on assumptions. If an extremely high ore price is tolerable, then very low grades of uranium ore can be considered as possible reserves. The DTI cites OECD/NEA 'Red Book' figures to claim that based on 2004 generation levels, known uranium reserves (at \$130/kg) will last for around 85 years (see References [2, 3]).

1.1.4 Energy Security and Disparity of Use

Energy security is a major concern worldwide. A large part of the world's oil is located in the Middle East and other politically unstable countries. The conflict between 'Western' and 'Islamic' cultures is at present exacerbating the anxiety over reliability of energy supply. Russia is a major producer of gas but recent events in Ukraine have made European countries aware how dependent they are on this single source. Until 2010, the USA was the world's largest consumer of energy (at which point it was overtaken by China following years of significant economic growth there); it remains dependent on imported oil and gas, although much less now than in the past due to increased levels of domestic production of gas from fracking. With economic growth seen as being intrinsically linked to cheap fuel it is difficult to imagine political parties, in the USA or elsewhere, proposing policies that require voters to drastically curtail their consumption and therefore alter their lifestyle.

Another disturbing aspect is the disparity in consumption between rich and poor countries: the richest billion people on the planet consume over 50% of all energy, while the poorest billion consume around 4%. This is an added source of tension and of accusations that the developed countries are profligate in their use of energy. To excuse this high consumption on grounds of high industrial activity is simply wrong. Japan for example is the world's third largest economy but has a per capita energy consumption roughly half that of the US. Pricing has a lot to do with this; it is not surprising perhaps that the US has the lowest costs by far for diesel and petrol (IEA Key World Energy Statistics, 2016, data for 2016).

1.2 The Environmental Impact of Energy Use

1.2.1 The Problem

Fossil fuels have one thing in common: they all create carbon dioxide when burned. They are a key part of the Earth's long term carbon cycle, having been laid down in geological periods when the climate was tropical across much of the planet and atmospheric CO_2 concentrations were very high. This storing of carbon through the growth of plant matter, and its subsequent conversion to coal, oil, peat and gas dramatically reduced atmospheric CO_2 levels and played an important role in cooling the planet to temperatures that could support advanced life forms. The concern now is that by unlocking this stored carbon climate change is being driven in the other direction, with global warming the direct result of an excessive greenhouse effect.

Ice core samples indicate that the level of carbon dioxide in the atmosphere was more or less stable at 260 parts per million (ppm) over the last few thousand years up to the onset of the industrial revolution at the beginning of the nineteenth century. Subsequently, atmospheric CO_2 levels rose, slowly at first as a result of coal burning, but since the Second World War the release of CO_2 has accelerated reflecting the exploitation of a wider range of fossil fuels. Current CO_2 levels are near 400 ppm and rising fast as made clear in Figure 1.3, taken from NASA based on ice core samples and other data.

CO_2 is not the only pollutant created by fossil-fuelled generation: combustion in air comprising 78% nitrogen by volume inevitably produces nitrogen oxides, NO and NO_2 and N_2O, collectively known as NOx; and any sulfur content of the fuel results in SOx emissions. NOx and SOx together contribute to acid rain and, as a result, it is now

Figure 1.3 Historic atmospheric CO_2 concentration taken from NASA. Source: Courtesy of NASA [6].

common to reduce any SOx emissions from fossil-fuelled power stations through flue gas desulfurisation. The down side of this is reduced thermodynamic efficiency and some resulting increase in CO_2 emissions.

World coal reserves are substantial, but coal is a less attractive fuel from the point of view of CO_2 emissions and also much more disruptive to extract. The cheapest coal is from open cast mines and this process is immensely damaging to the environment. All forms of generation have some environmental impact and these are not in general reflected in the cost of electricity; because of this, these additional environmental costs are known as *externalities*.

Externalities are consequences of activity that are not normally a part of the economic analysis. For example, the cost to society of ill health or environmental damage arising from pollution caused by a specific generating plant is not directly charged to the operator; that is, it is external to the micro-economics of the plant's operation. A number of European countries now seek to bring these externalities back into the economics of electricity generation by some kind of environmental levy or carbon tax. Carbon trading, discussed in detail in Chapter 7, is an alternative means of achieving this goal.

The nuclear cycle is of course not without externalities, although the environmental costs are highly contested, contributing as they do to the economic attractions or otherwise of nuclear power. Radioactive waste disposal, radioactive emissions and final decommissioning and disposal of radioactive reactor components are rarely fully accounted for and thus fall to an extent into the category of externalities. There are also issues concerned with environmental damage associated with uranium mining, but, in this regard, it is similar to coal. If nuclear power is to mitigate global emissions, it is of vital importance to accurately assess how much CO_2 will be displaced by nuclear power. This is a topic fraught with controversy. A common benchmark is the well-established 386 g CO_2/kWh contributed by gas fuelled power stations. The emissions for nuclear power are quoted as 11–22 g CO_2/kWh by OECD, and 10–130 g CO_2/kWh by ISA, University of Sydney [4]. A recent review of the literature by the InterGovernmental Panel on Climate Change (IPCC) (see next) identified a diversity of published results and calculated the 50th percentile figure as 16 g CO_2/kWh for nuclear power [5]. For reference, their figure for gas power generation was 469 g CO_2/kWh. If the higher published figures for nuclear turn out to be valid, the contribution of nuclear power to CO_2 mitigation would be seriously compromised.

1.2.2 The Science

The science of climate change is very well-established and its primary goal is to understand the link between CO_2 and other greenhouse concentrations and temperature rise. Work in this area has been carried out by the *IPCC*, which was set up in 1988 by the World Meteorological Organisation and the United Nations Environment Programme. It now involves scientists from up to 193 countries.

Figure 1.4, taken from NASA (Data source: NASA's Goddard Institute for Space Studies (GISS). Credit: NASA/GISS), updates results available in Reference [6], and shows the changing average global temperature, from 1880 to 2015. The bold curve is the smoothed trend while the individual annual averages are shown as points. The temperatures are shown relative to the average over 1940–1970. The earth has warmed by 0.9 °C since this period, bringing the global temperature to the warmest level in over

Figure 1.4 Temperature rise shown as global land-ocean temperature index (NASA, 2018).

12 000 years. All 10 warmest years on record have occurred since 1990 and there is considerable physical and biological evidence confirming climate change. Most climate models indicate that a doubling of greenhouse gases since the pre-industrial period is very likely to result in a rise between 2 and 5 °C in global mean temperatures. This increased level is likely to be reached between 2030 and 2060. If no action is taken, concentrations would be more than treble pre-industrial levels by 2100 resulting in a warming of 3–10° according to the latest climate projections.

Although the relationship between CO_2 concentration, temperature change and undesirable climatic changes is very complex and thus hard to predict precisely, it is widely believed that the CO_2 concentrations have to be stabilised if damaging global warming is to be avoided.

The IPCC concluded as far back in 2001 [7] that there was strong evidence that most of the warming observed over the last 50 years is *anthropogenic* in that it is attributed to human activities. This was supported by the Joint Statement of Science Academies (2005) and a report from the US Climate Change Science Programme (2006). The IPCC's most recent Fifth Assessment Report (AR5) confirms this conclusion and suggests that damaging global warming is now almost unavoidable [8].

The basic evidence that provides confirmation that global warming is due to human-made factors was provided by a climate model developed at the Hadley Centre for Climate Prediction and Research in the UK, and was known back in 2007. Nothing has changed since then to alter this understanding. In Figure 1.5, the observed global temperature since the early 1900s is shown by the bold line. The climate model was driven over that period by natural factors such as output of the sun, changes in the optical depth of the atmosphere from volcanic emissions and the interactions between the atmosphere and oceans. The predictions of the model are shown by the fuzzy band. This clearly disagrees with the observations particularly since about 1970 when observed temperatures have risen by about 0.5 °C, but those simulated by natural factors have not changed in this manner.

If the climate model is now driven by natural factors as previously but in addition by man-made factors – change in greenhouse gas concentrations and sulfate particles – the

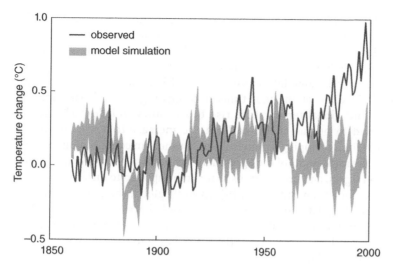

Figure 1.5 Natural factors cannot explain recent warming. Source: Crown Copyright 2007, the Met Office.

model simulation predictions in Figure 1.6 are in much better agreement with the temperature record. Climate modelling studies by other research centres have arrived at the same broad conclusions.

1.2.3 The Kyoto Protocol

The effects of climate change are global and hence mitigation requires coordinated international effort. Signed in 1997, the Kyoto Protocol aimed at reducing greenhouse gas emissions in the period 2008–2012 to 5.2% below those in 1990. The Kyoto Protocol

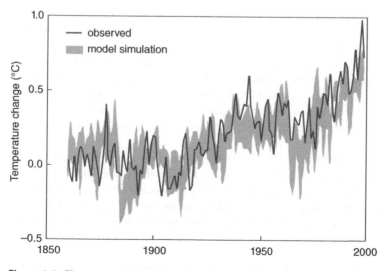

Figure 1.6 Climate warming can be correctly simulated when man-made factors are included. Source: Crown Copyright 2007, the Met Office.

is part of the United Nations Framework Convention on Climate Change (UNFCCC), established in 1994. The USA signed but did not ratify the Protocol and Canada withdrew from it in 2011. Emissions of greenhouse gases by the US are currently 20% higher than in 1990 while the target figure in Kyoto was a cut of 7%.

The Paris Agreement, which came into force in late 2016, is the most recent development within the UNFCCC. For the first time, it brings together all participating countries to work to limit global temperature rise this century to below 2 °C above pre-industrial levels, and to pursue effects to limit the increase to 1.5 °C. Financial support is to be given to the developing countries to assist them in contributing to these global aims. Although the aspirations of the agreement are admirable, the weakness is a lack of agreed national targets, and no sanctions that can be applied to ensure progress. According to the IPCC report on mitigation [9], to limit global temperature rise to less than 2 °C CO_2-equivalent (CO_2eq) concentrations would have to be limited to 450 parts per million (ppm) by the year 2100. This means that emissions must be between 40 and 70% below 2010 levels by 2050. In the IPCC guidance to policymakers within the mitigation report it is stated that

> Only a limited number of studies have explored scenarios that are *more likely than not* to bring temperature change back to below 1.5 °C by 2100 relative to pre-industrial levels; these scenarios bring atmospheric concentrations to below 430 ppm CO_2eq by 2100'

Given the present levels of CO_2-equivalent gases, this seems almost impossible.

It is clear, therefore, that it is prudent for industrialised countries to reduce their emissions by at least 60% by 2050 if the worst effects of climate change are to be mitigated with any confidence, and by more than this if the aspirational 1.5 °C limit to be achieved. This is a momentous challenge, to individuals, to governments and to supranational bodies.

Greatest responsibility rests of course with the nations producing the largest CO_2 emissions per capita and those moving fast up the emissions table. Table 1.1 illustrates the variation across selected countries/groupings of energy related emissions (i.e. excluding emissions associated with changes in land use and forestry), emissions per head and how this is partly driven by the income per head. Total emissions from China have already surpassed those of the US, although are much lower per capita. Figure 1.7, taken from the most recent IPCC report (AR5), shows clearly that while emissions from the OECD countries have been relatively stable between 1970 and 2010, Asia has seen a rapid expansion of emissions over the same period, and more recent data, for example in [1], confirms this tendency.

The developments reported by the IPCC in AR5[3] are not encouraging. Despite the range of climate change mitigation policies adopted around the world, anthropogenic greenhouse gas emissions continue to rise, indeed they are accelerating. Over the years 2000–2010 worldwide emissions grew by 2.2% whereas the growth rate was only 1.3% over the previous 30 years. Figure 1.8 shows the trends in emissions broken down by source; it is clear that the global economic crisis of 2008 caused only a brief reduction.

---

3 Although IPCC AR5 was published 2013, the data are from some years earlier due to the time required to collect and analyse the required information. It is worth stressing that this is the most up to date IPPC report of this series, and thus used here. The next one will not appear until 2021.

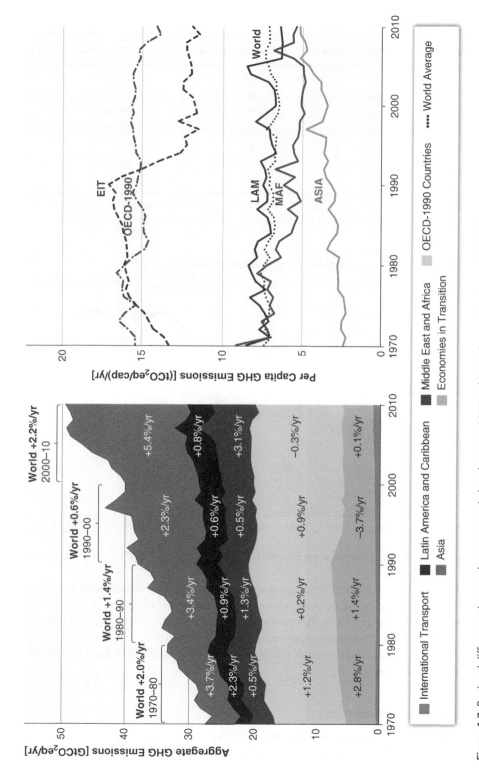

Figure 1.7 Regional differences in greenhouse gas emissions between 1970 and 2010. Source: Courtesy of IPCC [8].

Table 1.1 Greenhouse gas emissions for 2013, excluding land-use change and forestry.

Country/Grouping	GHG emissions MtCO₂ e	CO₂ per head (tCO₂)	GDP per head ($) 2015
USA	6280	19.9	56 115
EU	4224	8.3	32 005
UK	456	7.1	43 876
Japan	1353	10.7	34 524
China	11 735	8.6	8028
India	2909	2.3	1598
World	45 261	6.4	10 057

Taken from CAIT Climate Data Explorer. 2017. Washington, DC: World Resources Institute, available online at: http://cait.wri.org; population data from population reference bureau at http://prb.org/pdf13/2013-population-data-sheet_eng.pdf; and income per capita from the World Bank www.worldbank.org

Figure 1.8 Recent trend in greenhouse gas emissions. Source: Courtesy of IPCC [8].

AR5 also examines the target atmospheric greenhouse gas concentrations for 2100 and the associated likelihood of constraining global temperature rise to 2° relative to pre-industrial levels. It concludes that scenarios that reach 450 ppm CO_{2eq} by 2100 are more likely than not to limit temperature increase to less than 2 °C. On the other hand, scenarios that allow a concentration of 650 ppm or more are unlikely to limit the global temperature as required.

1.2.4 Economics of Mitigation

Alongside its work on climate change, the IPCC has studied the various mitigation options, and also the time dependent nature of the problem. To put it simply, the longer the delay in reducing greenhouse gas emissions, the more costly it will be to implement measures to stabilise the climate.

In the IPCC's AR5 report the effect of delay on mitigation costs is assessed through examination of large numbers of scenario modelling results, where the increase in costs is relative to a baseline optimal mitigation strategy. As shown in Figure 1.9, although there is a wide degree of uncertainty, the upward trends in costs are evident.

The economics of climate change mitigation were also addressed in the Stern Report, [10]. This work confirms that actions over the next 20–30 years will affect what happens in the decades to come. The Stern Report and the AR5 are in agreement that key actions should include:

- Increase in efficiency of energy use;
- Extensive use of renewable and other low-carbon technologies;
- Reduction in deforestation, indeed there should be reforestation.

Figure 1.10 from the Stern Report shows estimates of costs of low-carbon technologies in 2015, 2025 and 2050 that may be used to constrain CO_2 emissions. The costs are expressed as a central estimate, with a range, and as a percentage of the fossil-fuel alternative in the appropriate year. Due to learning effects the costs fall over time. The ranges reflect judgements about the probability distribution of unit costs and the variability of fossil-fuel prices. The 0% line indicates that the costs are the same as the corresponding fossil-fuel option. As expected, the uncertainties are large even for short term predictions. Onshore wind is shown to be particularly attractive with PV becoming very attractive beyond 2025.

On the basis of the costs of the low-carbon technologies and assumptions on possible rates of uptake over time, Stern's report estimates the distributions of emissions savings by technology for 2025 and 2050 for desirable climate stabilisation at 550 ppm. These estimates are shown in Figure 1.11. Energy efficiency and carbon capture and storage (CCS) play a major role in this scenario and will be discussed later in this chapter. Contributions from wind, solar, biofuels, hydro and distributed combined heat and power (CHP) through electricity generation provide the remaining savings and these are the technologies to be addressed in later chapters of this book.

1.2.5 Efficient Energy Use

Figure 1.11 stresses that efficiency measures are projected to make the largest contribution in climate change mitigation. It is therefore a surprise that the important topic of rational and efficient use of energy is rarely pursued vigorously in national or supranational plans in spite of the fact that study after study has shown that this route provides the most cost-effective way to meet sustainability goals.

In most countries, regulations and financial incentives are now in place to encourage energy efficiency but their effect is modest and national energy consumption figures continue to rise year on year. Energy efficiency must be the linchpin of any future energy strategy because [11]:

- Using energy as efficiently as possible is the most cost-effective way to manage energy demand, and thus to address carbon emissions. Saving energy is cheaper than making it.

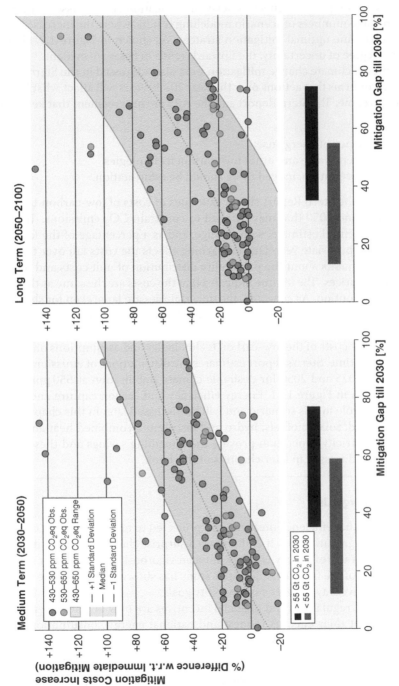

Figure 1.9 Cost impact of delays in implementing mitigation measures. Source: Courtesy of IPCC [8].

cost as % fossil fuel option

Figure 1.10 Unit costs of energy from low-carbon technologies CCS stands for carbon capture and storage, dCHP stands for distributed combined heat and power. Source: Courtesy of Cambridge University Press.

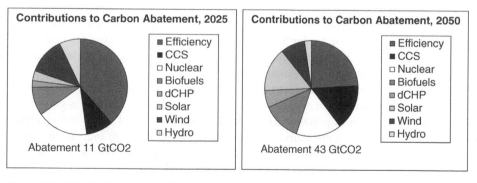

Figure 1.11 The distribution of emission savings by technology. Source: Courtesy of Cambridge University Press.

- By reducing demand on gas and electricity distribution networks, energy efficiency will improve the security and resilience of these networks and reduce dependence on imported fuels.
- By reducing energy bills, energy efficiency will help businesses to be more productive and competitive.
- Improving the energy standards of homes has an important role in reducing spending on fuel by those in fuel poverty.

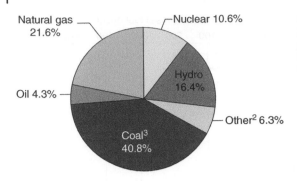

Figure 1.12 Contributions to the generation of electricity ([2]other includes geothermal, solar, wind, biofuels and waste, heat and so on[3]; coal includes peat and oil shale).

Increasing energy end use efficiency is unattractive for energy companies driven by commercial imperatives to increase sales and profits. It thus falls to governments to implement policies that change these drivers. Regulations can be put in place, for example, that require utilities to encourage customers to use electricity efficiently. A more revolutionary approach envisages the utility being transformed into a supplier of energy services, owning appliances in people's homes and thus being motivated to maximise the efficiency of these appliances. Whatever approach is finally adopted, the importance of reducing energy consumption should be the cornerstone of any CO_2 mitigation programme.

1.2.6 The Electricity Sector

Figure 1.12, taken from IEA 2016 statistics, [1], shows the percentage of fuels used in the generation of electricity. Fossil fuels account for 67% of the fuels used in this sector with coal being the dominant source at approximately 41% and contributing nearly three-quarters of CO_2 emissions. At present, large hydropower accounts for the major part of the renewables sector. Under half of the electricity produced is used in buildings, about a third in industry, under one-tenth in energy production (e.g. refineries) and less than one-tenth in transmission and distribution.

The world annual generation of electricity is in the region of 24 000 TWh representing an average rate of consumption of around 2500 GW. This electrical energy is generated in a very large collection of power stations driven mostly by fossil fuels. The electricity sector is the fastest growing source of emissions and estimated to increase fourfold between now and 2050. According to Stern, this sector would need to be at least 60% decarbonised by 2050 for atmospheric concentration to stabilise at 550 ppm and thereby reduce the risk of catastrophic climate change. Note that, as discussed previously, the more recent work by the IPCC indicates that 550 ppm is far too high. Consequently, much higher levels of decarbonisation will be required.

Figure 1.13, taken from Chapter 7 of the AR5 IPCC report, shows greenhouse gas emissions associated with the different forms of electricity generation. Emissions are divided into direct emissions (i.e. through fossil-fuel combustion), indirect emissions (for example in the production of the technology or operations and maintenance) and methane (reflecting fugitive losses associated with the process). The results emphasise the need to move from fossil fuels towards renewable sources. Note that, even with CCS, significant emissions remain for fossil-fuelled generation.

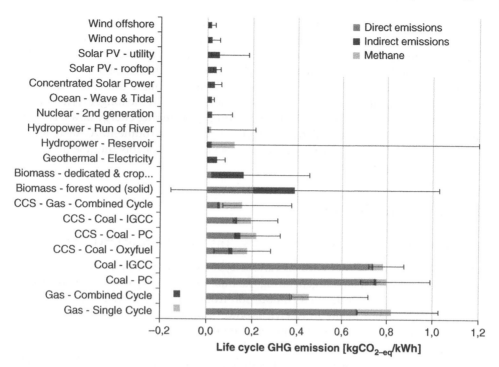

Figure 1.13 Life cycle emissions from the different sources of electricity generation.

1.2.7 Possible Solutions and Sustainability

Fundamental choices will have to be made in the years ahead. Our society is presently dependent on high and growing fossil-fuel consumption. The possibility of weaning ourselves from this dependence over a short time scale is completely unrealistic. The general shift in fossil fuels over the last two decades for both electricity generation and heating has been towards increased use of gas in place of coal and to a lesser extent oil. This has helped to limit the growth in CO_2 emissions as gas combustion releases less CO_2 per unit of energy than coal. Political events overseas, however, have generated anxiety in relation to increased dependence on this particular fuel.

A possible alternative path is to revert to dependence on coal. This resource is abundantly available in many developed counties including the US, Australia, many EU countries, Canada, Russia and in developing countries such as China, South Africa and Turkey. Recent developments in CO_2 capture and 'sequestration' for fossil fuels, discussed later in this chapter, gives some hope that this source may be made more acceptable environmentally.

There exist a number of potentially carbon neutral sources: these include nuclear fission (and possibly fusion in the far future) and all sources that derive directly or indirectly from the sun, namely biomass, wind, solar (thermal and photovoltaic), hydroelectric and marine. Geothermal and tidal energy are also carbon neutral and often regarded as renewable on the grounds that the sources are so huge as to be virtually inexhaustible. In Chapter 2, the characteristics of all these conventional and emerging technologies are discussed in some detail.

Finally, but no less important, other approaches essential in the move towards sustainability are a reduction in energy needs and improvements in the efficiency of energy use. The latter includes more efficient electricity generation. Although not the main focus of this book, these topics are briefly discussed in this chapter.

The planet's reserves of fossil fuels and minerals are of course finite, and thus the exploitation of coal, oil, gas and uranium are not sustainable in the longer term. Fortunately, renewable energy, being derived from naturally occurring energy flows, is inexhaustible and has no long-term detrimental effect on the environment. As such, it will in time become the basis of our energy supply system, and probably the sole means by which electricity is generated.

1.3 Generating Electricity

1.3.1 Conversion from Other Energy Forms – The Importance of Efficiency

Figure 1.14 shows the ways in which various types of energy can be converted into electricity. At present, the path generating the bulk of electricity world-wide is shown by the bold lines that lead through combustion from chemical to thermal, from thermal to mechanical and finally to electrical power conversion. The bottleneck of this path is the limited thermodynamic efficiency determined by the Carnot cycle. Older thermal generating stations have efficiencies between 35 and 40%, although in the last two decades conversion has been substantially improved to over 50% through the development of combined cycle gas turbines (CCGTs), a technology discussed in Chapter 2. It follows that when coal, oil or gas is used only 35–50% of the primary energy is successfully converted, the remaining being discharged into the environment in the form of waste heat.

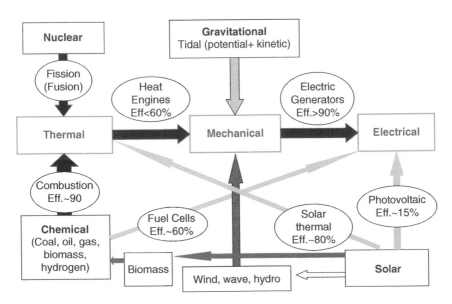

Figure 1.14 Conversion from a variety of energy forms into electricity.

One way of avoiding the Carnot limit is simply to make use of the waste heat. This is the principle of combined heat and power (CHP), used extensively in Scandinavia and of growing importance elsewhere. In such schemes, the waste energy from the thermal generation of electricity is distributed through heat mains to local industry and/or housing. This requires substantial infrastructure and is therefore only viable if the power station is reasonably close to the heat users. An alternative arrangement made possible by recent developments is to transport the fuel (mainly gas) to the consumer using the existing supply infrastructure and install the CHP system at the consumer's premises. Such systems are known as *micro-CHP* and are discussed in Chapter 8.

Direct paths that bypass the Carnot bottleneck are also available. The leading example of this approach is the fuel cell, which now borders on commercial viability in a number of forms: solid oxide, molten carbonate and PEM, to name the main ones. Another direct path is through photovoltaics, a technology that perhaps is the most promising in the medium term.

The conversion efficiencies of the various routes indicated in Figure 1.14 are dealt with in greater detail in Chapter 2.

1.3.2 The Nuclear Path

The topic of electricity generation from nuclear power elicits strong emotions from supporters and critics. Although nuclear power supplies only the equivalent of 5% of the world primary energy at the time of the writing of this book, some believe this should be expanded massively. They argue that it is an attractive source of electricity having very low-carbon emissions.

After the three-mile island and the Chernobyl accidents, there was a period of nearly 10 years during which almost no new nuclear capacity was constructed. Recently, the response to the aftermath of the Fukushima disaster was to say the least contradictory. Germany took the decision to close down all nuclear plant while the UK is welcoming a revival as the only way to satisfy its CO_2 reduction pledge. Additionally, recent concerns regarding fossil-fuel security have prompted a number of countries to consider new build programmes. China and India are planning to build several tens of reactors each and the USA is posed to do the same. In contrast, within Europe only Finland has embarked on the construction of a new nuclear plant, while Sweden, Switzerland and, as mentioned earlier, Germany all have moratoriums in place leading to a phasing out of nuclear power. France on the other hand remains committed to nuclear power, which contributes about 75% of its present electricity needs.

In the UK, the 2003 Government white paper was critical of the nuclear option, but by 2006, with concerns about a possible energy gap, the Government's position had changed. It is now actively supporting a new nuclear programme. A key concern is that major investments in nuclear will deprive renewable energy sources of the finance they need to expand. Reflecting its importance, the debate over nuclear power is extensive and there is voluminous literature. Reference [12] provides a good entry point for those interested.

1.3.3 Carbon Capture and Storage (CCS)

Figure 1.10 indicates that according to Stern, by 2025 and 2050 about 20 and 40%, respectively, of carbon abatement is expected to be provided by the emerging technology of

Geological Storage Options for CO_2

Figure 1.15 Geological storage options for CO_2. Source: World Coal Institute.

CCS provided that the techno-economic and environmental issues can be satisfactorily dealt with. CCS has the significant advantage of reconciling the necessary use of fossil fuels in the medium term with the necessity of serious cuts in CO_2 emissions.

A large-scale demonstration project was being planned in northeast Scotland, a joint venture by BP, Shell and ConocoPhilips [13]. Unfortunately, this project was abandoned but it graphically demonstrates the technology that is being vigorously pursued in other projects worldwide. Figure 1.15 illustrates the geological options for CO_2 storage.

With gas and oil prices likely to rise significantly, extracting CO_2 free energy from coal is also attracting substantial attention. Such technologies take a number of forms, but the co called Integrated Gasification Combined Cycle (IGCC) process is at the forefront. This involves the production of a synthetic gas (syngas) obtained from coal through gasification. Syngas is composed mainly of hydrogen and carbon monoxide and is the fuel source to a high-efficiency plant operating in combined cycle mode. Buggenum, a 253 MW plant in the Netherlands, operates on this principle and is the cleanest coal-based plant in Europe. To date, no IGCC plant involving carbon capture has been built although this principle was to have been used for a zero-carbon emission 275 MW plant funded by the US Department of Energy; this FutureGen project was abandoned in 2016 despite obtaining approval for a 30 mile pipeline to the injection site and draft permits for carbon sequestration.

1.3.4 Renewables

Figure 1.16 provides an overview of the earth's main energy paths that can be tapped to generate sustainable electricity. The main source of easily accessible renewable energy

Figure 1.16 Renewable energy flow paths.

is the sun. On average, the rate of solar radiation intercepted by the earth's surface is about 8000 times as large as the average rate of world primary energy consumption. With the present world population this amounts to a staggering average power of 2 MW per person.

The figure shows that this energy flux can be accessed directly using solar thermal or photovoltaic technology, or indirectly in the form of wind, wave, hydro and biofuels. Two other energy sources are often regarded as renewable in view of their sustainable nature: energy in the tides caused by the gravitational fields of the moon and the sun, which can be tapped using tidal barrages or tidal stream technology; and geothermal energy from the earth's core accessible in some locations through hot springs, geezers or boreholes. The available average power from these resources is a small fraction of that available from the sun.

A substantial proportion of the incident radiation is reflected back to space. Over the last several millennia and up to the onset of the industrial revolution, energy inputs and outputs have been in equilibrium at a global temperature level suitable to the development of the earth's biosphere. Exploiting the incident energy through the application of renewable energy technology *does not* disturb this balance. Intercepted natural energy flows, for example converted to electricity, and then converted again by consumers into mechanical, chemical or light energy, all eventually degrade into heat.

Most renewable energy forms are readily converted to electricity. Solar energy, geothermal energy and biomass can also be used to supply heat. Renewable energy can, in principle, provide all the energy services available from conventional energy sources: heating, cooling, electricity and, albeit with some difficulty and cost, transport fuels. It has the additional advantage that being a naturally distributed resource; it can also provide energy to remote areas without the need for extensive energy transport systems. It is worth noting that it is not always necessary to convert the renewable energy into electricity. Solar water heating and wind-powered water pumping are fine examples of systems that can work very well without involving electricity at all. However, the major contribution that renewable energy will be increasingly making in supplying our needs will be in electrical form.

Renewable energy is currently experiencing dramatic growth. Wind power and solar PV are leading the growth with global companies such as GE and Siemens now playing a leading role in the wind energy market, and China now taking the lead in PV with companies like Suntech and Yingli Solar alongside First Solar in the USA and Sharpe in Japan. In China, five of the largest electrical aerospace and power generation equipment companies have recently begun to develop wind turbine technology and Goldwind, Guodian and Sinovel all now have a significant presence in the market. Most large oil companies have expanded their research and development in Ethanol and Biodiesel production from biomass. The fastest growing RE technology is currently grid-connected PV with 40% annual year on year growth, but the RE technology that has made the largest contribution to date (excluding conventional hydro) is wind power with worldwide installed capacity approaching 580 GW by mid-2019; of this 184 GW is installed in the EU countries, about 117 GW in the USA and over 195 GW in China.

At least 48 countries have national targets for RE supply including all 28 EU countries. For the EU, binding targets for the renewable share of national total energy are set out in the Renewables Directive 2009/28.EC. The aim is to have 20% of all EU energy coming from renewable energy sources by 2020. Individual member states negotiated binding targets depending on their progress with renewable energy deployment and future opportunities. Figure 1.17 shows these targets and the status in 2012 taken from a Eurostat update from March 2014 [14].

1.4 The Electrical Power System

1.4.1 Structure of the Electrical Power System

Electricity is widely used because it is a supremely flexible form of energy. It can be readily and efficiently transported and is easily converted to other forms of energy. Mechanical energy can be provided by very efficient motors, light energy by increasingly efficient light fittings, heat energy by 100% efficient resistive elements and power supply to electronic and IT hardware through very efficient power conditioning units.

Figure 1.18 shows a diagrammatic layout of a typical electrical power system from the point of generation to the point of consumption. The figure depicts a coal fired power station as this, at present, represents the majority of generation plant.

The energy conversion chain follows the chemical→thermal→mechanical →electrical path depicted in Figure 1.14. Coal is pulverised and fed into a boiler where it is mixed with forced air and combusted. The boiler is a complex structure consisting of many stages of energy extraction from the combusted fuel. The flue gases are guided through equipment that remove solid particles and sulfur (de-sulfurisation not shown in the figure) before being released into the atmosphere. The highly purified water in the boiler is converted into superheated steam that is passed through several turbine stages on the shaft of a turbo-generator. The low-pressure low temperature steam from the outlet of

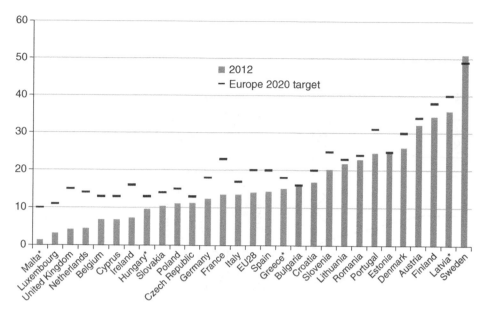

* estimated

Figure 1.17 Progress towards EU Renewables Directive targets.

the turbine is condensed into the purified water that in this closed system is pumped back into the boiler. The condensing process unfortunately needs substantial amount of external cooling water. In the figure, this water is provided from a pond at the bottom of a cooling tower. The hot water from the condenser is sprayed at the top of the tower and transfers its heat to the air that passes up the venturi shaped tower. The lost water must be made up from some external source such as a local river.

The energy generated at the power stations is transmitted to the consumers using overhead transmission lines and underground cables that possess ohmic resistance. The energy loss due to the unavoidable resistive heating of a line or cable is proportional to the square of the current I it carries. Additionally, for a given power transfer that is proportional to VI, the current is inversely proportional to the voltage (other things being equal). Thus, the loss decreases with the square of the voltage V. The downside of operating at higher voltages is that costs of insulation and other power system equipment increase substantially. Thus, for bulk transmission of power over long distances, higher voltages are most economic whereas, for local distribution of modest power to numerous connection points, lower voltages are most economic. The economics also dictate that it is worthwhile having several intermediate voltages. This multiple voltage arrangement results in network transmission losses confined to within 5–10% of the throughput power.

The bulk of global electricity is generated in large (>500 MW) power stations at around 20 kV. This is then stepped up by transformers to an *extra high-voltage* (EHV) level such as 400 kV and carried by the transmission system to the bulk supply points, where it is stepped down to a *high-voltage* (HV) level of around 100 kV. Some very large industrial consumers are connected at this level but most power is transformed down

Figure 1.18 Pictorial view of the components of a large power system.

again to *medium voltage* (MV) levels such as 30 kV, then to 10 kV and finally to the *low voltage* (LV) level of 400 V also referred to as the *distribution system*, which provides 230 V when the connection is single phase. In the USA and few other countries, the LV level is 200 V three-phase, 115 V single phase. The voltages used vary from country to country but the power system structure follows closely the layout of Figure 1.18.

In Figure 1.18, a *circuit-breaker* is shown after the generator step-up transformer. This is a component part of an extensive protection network that permeates all levels of the power system. Faults on the network may result in low resistance paths that cause excessive currents capable of damaging equipment. The protection devices, circuit breakers at HV levels and *fuses* at domestic distribution level, operate to isolate the faulty part of the network and prevent equipment damage. The effect on the protection system of introducing renewable energy sources will be discussed in Chapter 6.

1.4.2 Integrating Renewables into Power Systems

The term *grid* is often used loosely to describe the totality of the network. In particular, *grid-connected* means connected to any part of the network. The term *national grid* usually means the EHV transmission network.

Integration specifically means the physical connection of the generator to the network with due regard to the secure and safe operation of the system and the control of the generator so that the energy resource is exploited optimally. The proper integration of any electrical generator into an electrical power system requires knowledge of the well-established principles of electrical engineering. The integration of generators powered from renewable energy sources is fundamentally similar to that of fossil-fuelled powered generators and is based on the same principles, but renewable energy sources are often time variable and geographically dispersed.

A renewable energy generator may be described either as *stand-alone* or *grid connected*. In a stand-alone system a renewable energy generator (with or without other back-up generators or storage) supplies the greater part of the demand. In a grid-connected system, the renewable energy generator feeds power to a large interconnected grid also fed by a variety of other generators. The crucial distinction here is that the power injected by the renewable energy generator is only a small fraction of that generated by the totality of generators on the grid. The distinction between stand-alone and grid-connected generators is a useful one but is not always clear-cut. Sometimes confusion arises when the word *grid* is used to refer to a relatively small stand-alone electrical network. This is not necessarily wrong (although it may indicate delusions of grandeur!) but it should always be made clear as to the extent of the grid being referred to.

The point on the network to which a renewable energy generator is connected is referred to, for reasons to be explained later, as the *point of common coupling* (PCC).

1.4.3 Distributed Generation

Power systems have developed over the years to supply a varying demand from centralised generation sourced from fossil and nuclear fuels. Unless nuclear fusion proves successful, which won't be known for over 50 years, there is universal agreement that by the end of this century the majority of our electrical energy will be supplied from RE sources.

Generators powered from renewable energy sources (except large-scale hydro large offshore and onshore wind farms and increasingly large photovoltaic installations) are typically much smaller than the fossil-fuelled and nuclear-powered generators that dominate today's large power systems. Small generators cannot be connected to the transmission system because of the cost of high-voltage transformers and switchgear. Also, the transmission system is often a long way away as the geographical location of the generator is constrained by the geographical availability of the resource. Small generators must therefore be connected to the distribution network. Such generation is known as *distributed* or *dispersed generation*. It is also known as *embedded generation* as it is *embedded* in the distribution network.

In traditional power systems power invariably flows from the large centralised power stations connected to the EHV network *down* through the HV and LV systems to be distributed to consumers. In power systems with distributed generation, power may travel from point to point within the distribution system. This unusual flow pattern has some serious implications in the effective operation and protection of distribution network. Distributed generation will be discussed in Chapter 6.

It may be concluded that present power systems will gradually have to evolve and adapt so that in the far future, a managed demand will be supplied from distributed, mostly time variable, RE generation. This transformation will be aided by the liberal use of power electronic interfaces capable of maximising the effectiveness of RE sources, controlling power flows and ensuring reliability of supply. Some of these issues are discussed in Chapter 8.

1.4.4 Renewable Energy Penetration

The proportion of electrical energy or power being supplied from renewable sources is generally referred to as the *penetration*. It is usually expressed as a percentage. When fuel or CO_2-emission savings are being considered, it is useful to consider the *average penetration*:

$$Average\ Penetration = \frac{Energy\ from\ renewable\ energy\ powered\ generators\ (kWh)}{Total\ energy\ delivered\ to\ loads\ (kWh)}$$

In this case, the energy (kWh) is measured over a long period of time, perhaps a year. At first sight, it might seem more natural to express the denominator (*total energy delivered to loads*) as: *total energy from all the generators (including fossil-fuelled generators)*. However, in stand-alone systems, there may be *dump loads* (loads where energy is dumped as heat) to consider, and, in grid-connected systems, there is often interest in the penetration in a given geographical area, in which case it may be termed the *local penetration*.

For other purposes, including system control, it is necessary to consider the *instantaneous penetration*:

$$Instantaneous\ Penetration = \frac{Power\ from\ renewable\ energy\ powered\ generators\ (kW)}{Total\ power\ delivered\ to\ loads\ (kW)}$$

Since the electrical output from most generators operating from renewable energy sources is time variable, the *maximum instantaneous penetration* will normally be much higher than the average penetration.

1.4.5   Network Stability

As renewable energy penetrations inevitably rise as countries around the world move to reduce their CO_2 emissions, there will be rising concern about network stability. Engineers can mean rather different things when they talk about system stability. *Power system stability* can be defined as the ability of the system to quickly return to a stable operating condition after being exposed to a disturbance such as, for example, the loss of a large generating station, or the loss of an overhead transmission due to short circuit or some other fault resulting in the automatic disconnection of that line by its protection systems. In engineering terms, a power system is deemed stable if the rotational speeds of motors and generators, and substation voltage levels must return to their normal values in a quick and stable manner. Transmission and distribution engineers tend to put emphasis on voltage, and indeed some talk about voltage stability. And there are different and complex ways in which power systems can become unstable including inter-area modes where different geographically located parts of the power system become involved in potentially unstable interactions. These and related issues will be discussed in more detail in Chapter 5. Non-technical readers can at best hope initially to gain a qualitative understanding of these issues that can be mathematically difficult. The topic has been added since the first edition because of its increasing importance and the rising concern among network operators, especially of the smaller power systems such as those of Britain, Ireland and Japan.

References

1 IEA, Key World Energy Statistics, International Energy Agency, 2016.
2 Barnaby F., and Kemp J. 'Too hot to handle? The future of civil nuclear power' Briefing paper, July 2007, Oxford Research Group.
3 Cited in 'The future of nuclear power' (DTI/Pub 8519/4K/05/07.NP): p. 156: available at www. http://dti.gov.uk/files/file_39197.pdf (accessed 2008).
4 Australian Government. Available at: www. http://pmc.gov.au/umpner/docs/commissioned/ISA_report.pdf (accessed 2008).
5 Moomaw W., Burgherr P., Heath G., Lenzen M., Nyboer J., Verbruggen A., Annex II: Methodology. In IPCC: 'Special Report on Renewable Energy Sources and Climate Change Mitigation' (p. 10), 2011.
6 Morice, C.P., Kennedy, J.J., Rayner, N.A., and Jones, P.D. (2012). Quantifying uncertainties in global and regional temperature change using an ensemble of observational estimates: the HadCRUT4 dataset. *J. Geophys. Res.* 117: D08101.
7 IPCC, 2001: Climate Change 2001: Synthesis Report. Contribution of Working Groups I, II and III to the Third Assessment Report of the Intergovernmental Panel on Climate Change. (Watson R.T. and the Core Writing Team, eds). Cambridge University Press, Cambridge, United Kingdom and New York, NY, USA, 2001.
8 IPCC, 2014: Climate Change 2014: Synthesis Report – Summary for Policy Makers. Contribution of Working Groups I, II and III to the Fifth Assessment Report of the Intergovernmental Panel on Climate Change (Core Writing Team, Pachauri R.K. and L.A. Meyer, eds). IPCC, Geneva, Switzerland, 151 p.

9 IPCC, 2014: Climate Change 2014: Mitigation of Climate Change. Contribution of Working Group III to the Fifth Assessment Report of the Intergovernmental Panel on Climate Change (Edenhofer O., R. Pichs-Madruga, Y. Sokona, E. Farahani, S. Kadner, K. Seyboth, A. Adler, I. Baum, S. Brunner, P. Eickemeier, B. Kriemann, J. Savolainen, S. Schlömer, C. von Stechow, T. Zwickel and J.C. Minx, eds). Cambridge University Press, Cambridge, United Kingdom and New York, NY, USA, 2014.

10 Stern Review Report on the Economics of Climate Change. Cambridge University Press, 2006 Following criticisms of the report a rebuttal was published by Stern and his team and can be found in: http://www.hm-treasury.gov.uk/media/5E1/FB/stern_reply_worldeconomics.pdf, 2006.

11 Parliament UK, Plan for Action on Energy Efficiency. UK Government follow up to the Energy White Paper, April 2004.

12 Sustainable Development Commission (2006). *The Role of Nuclear Power in a Low Carbon Economy*. UK: SDC.

13 Gordon S. Carbon cure. IEE Power Engineering, Dec./Jan. 2005/2006.

14 Eurostat News Release 37/2014, 10 March 2014.

2

Features of Conventional and Renewable Generation

2.1 Introduction

The purpose of this chapter is to describe the essential features of the different electricity generation plant so that the role of the renewable sources and how they might be integrated into the electricity supply system can be better understood. The main features of conventional sources are dealt with in a single section and in less detail since there are many reference works available covering this ground. Moreover, although such sources supply the overwhelming proportion of energy in power systems into which renewables are now being integrated, they are themselves not the focus of this book. Such plant will be with us for a long time to come and in the transition to a sustainable supply system they will need to complement the growing proportion generated from renewable sources, and as time progresses this plant will be required to operate more flexibly. It is therefore essential that the reader has a basic understanding of conventional sources and their characteristics. Each of the renewable energy sources is treated in greater length and separately to reflect their rather diverse characteristics.

First, it should be noted that, although perhaps rather obvious, the law of conservation of energy requires that the time varying power demand of the consumers must be continuously matched by the generated power. In this situation, generation is said to be *load following*. How this is done and what happens when exact balance is not achieved is the topic of the next chapter. Conventional sources, taken here to include both fossil-fuelled and nuclear generation, in general produce a given output when required to by the operator and so can be *dispatched*,[1] whereas most renewable sources generate according to the time varying strength of the renewable energy source. These time variations are particularly noticeable in the case of wind, solar, wave and tidal, although different in the extent to which the variations can be predicted, a factor that will turn out to be significant for efficient operation of the power system. Conventional generation in itself is neither instantly dispatchable nor completely flexible; it takes time to prepare large thermal plant for full load output and such plant are not permitted to operate at too low a load. Nuclear plant in particular offers very limited capability to follow changes in load.

The remainder of this chapter explores in detail the characteristics of the different sources of generation with specific regard to their contribution to meeting the demands of a large power system. All the renewables are dealt with except geothermal power, which is very geographically limited, and were it does exist can be treated very much

---

1 Dispatching is the planned allocation of plant to meet expected future loads on the system.

Renewable Energy in Power Systems, Second Edition. David Infield and Leon Freris.
© 2020 John Wiley & Sons Ltd. Published 2020 by John Wiley & Sons Ltd.
Companion website: www.wiley.com/go/infield-renewable-energy

like fossil-fuelled thermal power. It is, after all, simply a way of raising steam using large naturally occurring thermal gradients.

The importance of load forecasting will be explained in Chapter 3. When time variable renewable energy becomes significant in a power system, forecasting this also becomes important. For each of the renewable sources, the accuracy of forecasts will be discussed.

2.2   Conventional Sources: Coal, Gas and Nuclear

The diverse characteristics of generators operating from different fossil fuel sources define their function in an integrated power system. Nuclear-powered stations, dealt with in more detail in the next sub-section, are generally inflexible and designed to run at constant power. This mode of operation is known as *base load* since it contributes to that fraction of the load that can be regarded as always present.

Conventional coal plant[2] efficiency, as reviewed in Chapter 1, is in the range 30–40% and depends significantly on how it is operated. The higher efficiencies would correspond to the best plant in base load operation; increased *cycling*[3] and low load operation of the plant significantly lowers operational efficiency. Peak system loads generally occur for only short periods of time and it turns out that such loads are best met by plant of low capital cost[4]; the high operational costs associated with typically low efficiency generation of this peaking plant is acceptable since total operation time tends to be limited.

Thermal conversion has been substantially improved over the last two decades through the development of combined cycle gas turbines (CCGTs). In this arrangement the gas, usually natural gas, is burnt at temperatures of around 1000 °C to drive a turbine; the exhaust gases are subsequently used to raise steam for a traditional steam turbine. Both turbines drive separate generators feeding power to the grid. By increasing the temperature at which the fuel is burnt the efficiencies of around 50% can currently be achieved, with plant efficiencies towards 60% being projected for the next generation. Such plant are currently highly controllable, however, because of their high efficiency they are often constrained by fuel supply contracts that dictate running on high load. Most gas-powered stations are therefore run at constant power and thus also contribute to the base load, although it is worth noting that recent rises in gas prices within Europe have changed the situation to one where it is, at least for the time being, economically preferable to run coal-fired stations at base load in place of CCGT plant, and that emphasis is being given to the flexible operation of CCGTs.

The design of generation plant does affect its time response and the extent that it can contribute to the regulation demands of the power system. Load following capabilities reflect in the main the thermal mass of the central plant elements. Thus, coal stations with large boilers can take many hours to reach full output from cold. In contrast, gas turbines can reach rated power in minutes and combined cycle plants lie

---

2 Most conventional modern coal plant burn pulverised coal, future plant using gasified coal could reach considerably higher overall efficiencies.

3 Plant cycling is the process by which a plant is run up from cold to meet load and subsequently run down when not required. During the transitional periods, plant is generally operating at far from optimal conditions and energy is required to prepare the plant for generation. All this contributes to poor overall efficiency.

4 Open cycle gas turbines provide cheap peaking capacity but are relatively expensive to operate. The impact of plant costs on their role in the system is explored in more detail in Chapter 7.

somewhere between coal-fired and direct gas fired plant. Innovations in plant design aimed at improving conversion efficiency can have the effect of reducing flexibility. This important issue is discussed in detail in Reference [1].

Although the term nuclear power formally covers both fission and fusion, fusion power is still decades away and may never be technologically viable or cost effective, it certainly cannot be regarded as conventional. Nevertheless, considerable international research has been and continues to be directed towards this technology, in particular through the ITER consortium. In contrast, fission-based reactors have been generating a significant proportion of electricity world-wide. It remains a controversial technology due to the dangers of radiation and the challenge of radioactive waste disposal. A number of different technologies have been developed including boiling water reactors and gas cooled reactors.

Such plant sustain a fission chain reaction within a controlled environment. Fission takes place in the reactor core that is normally contained within a pressure vessel safety shield. Moderators, usually graphite or water, slow down the neutrons and help regulate the chain reaction. Control rods are inserted into the core to regulate the nuclear reaction; these are made of a material, such as boron, that absorbs neutrons. The core also comprises fuel rods of fissile material. A coolant, normally water or gas, passes through the reactor. It then passes to the boiler where steam is raised. Electrical generation is provided by steam turbines and in this regard nuclear power stations are similar to large coal-fired stations, it is simply the source of heat for raising the steam that is different.

In general, nuclear reactors are regarded as inflexible and they normally operate on base load. The fact that capital and installation costs far outweigh fuel and operational costs is further motivation for operating such plant as continuously as possible. Load factors are around 80% but would almost certainly fall if they were used to any extent to load follow as advocates of nuclear power in France have claimed is possible. The activity of load following is sometimes confused with the steady reduction of nuclear power station output over some weeks that is possible, indeed advisable, if the station is being run down prior to maintenance.

2.3   Hydroelectric Power

As mentioned in Chapter 1, hydroelectric generation is an indirect form of solar energy. Incident solar radiation evaporates water from the sea, and to a lesser extent from land areas and the warmed water vapour rises; as it ascends it expands and cools, eventually condensing in the form of clouds. Some of the resulting rain falls on high ground. This water has thus gained potential energy as a result of solar input. Hydro power is the result of extracting some of this energy as the water flows back towards the sea. Large-scale hydro makes use of large reservoirs, usually created by damming rivers. Water is allowed to flow out of the reservoir in a controlled manner, turning turbines that drive electrical generators as it does so.

The storage of water in the reservoir allows generation to be timed to meet the demands of the power system. Energy storage capacity[5] is limited so the aim is usually to generate at times of high load and so maximise the income generated. Since

---

5 Capacity for a store is the energy that can be stored (i.e. MWh), rather than the rating.

water availability is limited and seasonal, complex algorithms making use of rainfall prediction are used to optimise hydro operation. Some countries, for example Norway and Switzerland, have bountiful hydro resources and as a result there are times when electricity is so cheap it is almost given away. This has encouraged industries to develop that require abundant cheap electricity. Energy intensive processes such as aluminium smelting and silicon production are often located in countries with plentiful hydro resources for exactly this reason.

Dams are attractive because they can provide a large *head* (equal to the fall in height of the water) but whether they can be built depends on the local topography. Rivers are the natural way in which water loses potential energy and it is possible to extract a proportion of this by means of so called low-head schemes that are usually small-scale. It is also possible to place turbines in a river flow directly with no dams or penstocks[6], extracting only a very small amount of the predominantly kinetic energy as the river flows by. Such turbines operate at effectively zero head and the installations are known as run-of-river schemes.

2.3.1 Large-Scale Hydro

Large-scale hydro is a well-developed and widely used form of generation. Depending on estimates, between 20 and 25% of the world's large-scale hydro potential has already been developed, although the resource reasonably located geographically in relation to electricity demand has according to many commentators already been largely exploited. Hydroelectric stations currently contribute about 20% of world electricity generation. Large-scale hydro power is operationally the most desirable of all renewable energy sources with respect to availability and flexibility of supply.

As explained previously, water can be stored in reservoirs and used when required, either continuously if the reservoir is large or when most required by the demand for electricity. The advantage of this storage arrangement is compounded by the natural capability of hydro plant to respond within minutes to demand increase or decrease. Such plant is therefore invaluable as flexible generation to follow both predicted and unexpected changes in the consumer demand. This feature is so valuable that in countries such as the UK where the topography is unsuitable for the installation of large conventional hydro, *pumped-storage* schemes driven by the bi-directional transfer of water between two reservoirs have been developed.

The down-side of large hydro schemes is that they involve substantial up-front capital investments with profits accrued over long periods in the future. Their development can also be environmentally undesirable because of the flooding of large areas and the displacement of populations, such as has occurred with the construction of the Three Gorges scheme in China. However, this established renewable energy technology is unlikely to provide substantially increased contributions in the future since many of the attractive sites have already been developed; the areas with remaining potential lie mainly in the former Soviet Union and the developing countries, in particular Turkey. A good recent review can be found in Reference [2]. The optimal way of integrating this technology in power systems has been covered in many books and scientific papers published over the last century and will not be dealt with further here.

---

6 A penstock is sluice or gate used to control the water together with the pipe to take the water to the turbine.

Hydro scheduling methodology is well established (see e.g. Reference [3]), although more recent approaches based on machine learning are now available, such as in References [4, 5].

2.3.2 Small Hydro

Small-scale hydro is commonly defined as being smaller than 5 MW. At even lower powers (<100 kW), so called micro-hydro is subdivided into dammed and *run of the river* in which there is no storage; a further distinction is into high and low head. It is estimated that the world potential of small/micro-hydro is around 500 GW, of which roughly only one-fifth has been exploited to date. A particular attraction of small/micro-hydro is that the resource is often located in remote rural upland areas unserved by conventional electricity supply, but where the size of communities and their energy requirements are consistent with the available supply. The flow in a given river will vary greatly throughout the year, generally having, in the Northern hemisphere, high values during the winter months and low values during the summer months. In more tropical climates the flow is likely to relate to monsoon conditions. These seasonal constraints are important. For example, a correctly sized small-scale hydropower scheme in the UK will not be expected to have sufficient water to run continuously throughout the summer months. For this reason, most small hydro systems are grid connected.

Figure 2.1 shows the mean daily flow for the river Barle in Somerset for 1980 from January through to December. The following observations can be made:

- For two months in the summer, the flow dropped to very low values, less than 1 m³/s.
- The river flooded (over 20 m³/s for this river) on several occasions.
- The river level tends to rise fairly quickly but this level reduces gradually; that is, there is a sharp 'leading edge' followed by a slow decay. This is a general characteristic of rivers.

Small-scale hydro schemes show little variability in output from minute to minute, but can change substantially over hourly or daily cycles due to sudden rainfall. As already mentioned, the output is also highly sensitive to the time of year.

Figure 2.1 The mean daily flow on the river Barle at Dulverton, Somerset in 1980. Source: Courtesy of Robin Cotton of Loughborough University, 1998.

If a large number of small-scale hydro schemes are connected to an integrated electricity network their influence on the minute to minute operation of the power system would be negligible due to the statistical smoothing effect from the aggregation of uncorrelated short-term variations.

2.3.2.1 Turbine Designs

All hydro systems exploit the potential energy of the water. Engineers use the term head, H, for the height through which the water is allowed to fall, often in practice this becomes the net or effective head, reflecting frictional losses. Potential energy is simply mgH where g is the acceleration due to gravity; for a volume V of water density ρ this is $V\rho gH$. The power in kilowatts associated with a flow rate Q m^3/s falling through an effective head of H metres is thus given by:

$$P = Q\,\rho\,gH$$

Two different approaches to turbine design exist: impulse turbines like the Pelton wheel extract kinetic energy from the flow through impact on cups mounted on the turbine wheel; and reaction turbines like Francis and Kaplan designs, which in contrast to impulse turbines, operate submerged in water. Impulse turbines are suited to high pressure/head and can have efficiencies around 90%, whereas reaction turbines, run faster and are suited to lower heads and have higher efficiencies. Turbine efficiency is conventionally plotted, as in Figure 2.2, as a function of the *specific speed*, $N_S = n\sqrt{(P)}/H^{5/4}$ where P is the output power in kW, H is the effective head in metres and *n* is the rotational speed in revolutions per minute (rpm).

The formula for the specific speed can be understood by considering a turbine scaled down to produce 1 kW of power at a head of 1 m for then its runner will rotate at the specific speed equal to its speed in rpm; that is, $N_S = n$.

The speed of the runner blade relative to the water striking it is critical for the efficiency of the turbine. Propeller type turbines run best when their blade tips move faster than the water. In the case of the Kaplan, the blades should move at twice the water velocity, whereas the skirted Francis is most efficient when the two speeds are roughly equal. The Pelton theoretically performs best when its blades are moving at half the water velocity.

Reaction turbines use guide vanes at the rotor inlet to adjust the direction of the inlet flow and the flow rate so as to vary the shaft power in response to electrical load.

Figure 2.2 Efficiency curves for different turbine types.

The turbine runner has a series of vanes whose complicated geometry is designed to extract maximum power under design conditions. These vanes cannot be moved instantaneously and this has implications for the power quality achievable from such turbines. For stand-alone applications where fast control is essential, electronic control of the loads is used to maintain turbine speed.

From Figure 2.2, it is clear that efficiency depends critically on specific speed, and this in turn depends on the effective head; thus, if the effective head varies significantly, the speed of the rotor needs to be adjusted to maintain high efficiency. Most turbines, however, operate at a fixed speed and efficiencies may thus vary to some extent.

Small hydro systems can be grid connected by driving an appropriately sized induction generator (see Chapter 4) provided that there is sufficient flow control. Flows should be controlled by the guide vanes to limit the output of the generator to the rated value.

2.4 Wind Power

2.4.1 The Resource

Winds result from the large-scale movements of air masses in the atmosphere. These movements of air are created on a global scale primarily by differential solar heating of the earth's atmosphere so wind energy, like hydro, is also an indirect form of solar energy. Air in the equatorial regions is heated more strongly than at other latitudes, causing it to become lighter and less dense. This warm air rises to high altitudes and then flows northward and southward towards the poles where the air near the surface is cooler. This movement ceases at about 30 °N and 30 °S, where the air begins to cool and sink and a return flow of this cooler air takes place in the lowest layers of the atmosphere.

The areas of the globe where air is descending are zones of high pressure. Conversely where air is ascending, low pressure zones are formed. This horizontal pressure gradient drives the flow of air from high to low pressure, which determines the speed and initial direction of wind motion. The greater the pressure gradient, the greater is the force on the air and the higher is the wind speed. Since the direction of the force is from higher to lower pressure, the initial tendency of the wind is to flow perpendicular to the isobars (lines of equal pressure). However, as soon as wind motion is established, a deflective force is produced due to the rotation of the earth that alters the direction of motion. This force is known as the Coriolis force. It is important in many of the world's windy areas, but plays little role near to the equator.

In addition to the main global wind systems there are also a variety of local effects. Differential heating of the sea and land also causes changes to the general flow. The nature of the terrain, ranging from mountains and valleys to more local obstacles such as buildings and trees, also has an important effect.

The boundary layer refers to the lower region of the atmosphere where the wind speed is retarded by frictional forces on the earth's surface. As a result, wind speed increases with height; this is true up to the height of the boundary layer, which is at approximately 1000 m, but depends on atmospheric conditions. The change of wind speed with height is known as the wind shear.

It is clear from this that the available resource depends on the hub height of the turbine; this has increased over recent years, reflecting the scaling up of wind turbine technology with the hub heights of multi-megawatt machines now up to 150 m.

The European accessible onshore wind resource has been estimated at 4800 TWh/year taking into account typical wind turbine conversion efficiencies, with the European off-shore resource in the region of 3000 TWh/year although this is highly dependent on the assumed allowable distance from shore. A recent report suggests that by 2030 the EU could be generating 778 TWh a year from a projected onshore capacity of 254 GW with 66 GW offshore, amounting to 24.4% of forecast EU total electricity requirements, (Reference [6]). The world onshore resource is approximately 53 000 TWh/year taking into account siting constraints. To see these figures in context, note that the UK annual electricity demand is in the region of 350 TWh and the USA demand 3500 TWh. No figure is currently available for the world offshore resource, and this itself will be highly dependent on the allowable distance from shore.

Of the new renewables wind power is the most developed. On very windy sites wind farms can produce energy at costs comparable to those of the most economic traditional generators. Due to advances in technology, the economies of scale, mass production and accumulated experience, over the next decade wind power is the renewable energy form likely to make the greatest contribution to electricity production. As a consequence, more work has been carried out on the integration of this resource than any of the other renewables and naturally, this is reflected in the amount of attention given to wind power integration in this book. Solar power (mainly photovoltaics) is likely to be of increasing importance over the next decade and research into its integration into power systems is becoming more common; key issues will be dealt with in later chapters.

2.4.2 Wind Variability

The wind speed at a given location is continuously varying. There are changes in the annual mean wind speed from year to year (*annual*), changes with season (*seasonal*), with passing weather systems (*synoptic*), on a daily basis (*diurnal*) and from second to second (*turbulence*). All these changes, on their different time scales, can cause problems in predicting the overall energy capture from a site (annual and seasonal), and in ensuring that the variability of energy production does not adversely affect the local electricity network to which the wind turbine is connected.

Wind characteristics will vary considerably from site to site, with more elevated and exposed sites having higher annual mean wind speeds. Turbulence is short-term wind speed variation mechanically generated when the wind flows over terrain and is related to the roughness of the surface and height above the ground. Sites are characterised by a surface *roughness length*, denoted z_0, measured in metres. This varies from about 10^{-4} m for very flat and open sites or calm seas, to 10^{-1} m for grass and scrubland, to over 1 m for forested or urban sites. The surface roughness also affects the wind shear, which is the increase in wind speed with height.

For high winds overland there exists a simple expression known as the log law that relates the wind speed at the turbine hub height z, denoted $U(z)$, to the wind speed at a lower, possibly measurement, height z_r:

$$U(z) = U(z_r) \left[\frac{ln(z/z_0)}{ln(z_r/z_0)} \right] \tag{2.1}$$

Technically, *turbulence intensity* is defined as the ratio of the short-term wind variation as measured by the standard deviation of rapidly sampled wind speed over

10 minutes, divided by the mean wind over that time. It is conventionally denoted by I and is often stated as a percentage. For an onshore wind site, 10% would be a low value, more typical of conditions offshore, while rugged onshore sites with over 25% turbulence can be encountered. Developers try to avoid high turbulence due to the higher rates of fatigue damage that will occur. A useful approximation for the variation of turbulence intensity with height is:

$$I(z) \approx \frac{1}{ln(z/z_o)} \tag{2.2}$$

In Figure 2.3, each graph shows the wind speed over the time periods indicated. Wind speed measured continuously over 100 days is shown on the first graph followed by graphs that in sequence zoom in on smaller and smaller windows of the series. It is easy to see the much larger relative variability in the longer time series (synoptic) as compared with the time series covering hours or less (diurnal, turbulence). This information is summarised in the spectral density presentation in Figure 2.4. In a spectral density function the height indicates the contribution to variation (strictly the variance) for the frequency indicated. A logarithmic scale as used here is the norm, and allows a very wide range of frequencies/time scales to be represented easily. The y-axis is scaled by n to preserve the connection between areas under any part of the curve and the variance. The area under the entire curve is the total variance. It can be seen that the largest contribution to variation is the synoptic variation, confirming the interpretation of Figure 2.3.

Figure 2.3 Wind speed measured 30 m above flat terrain (Vertical axis is wind speed, 0–20 m s^{-1}). Source: Reproduced with permission of Riso National Laboratory for Sustainable Energy.

Figure 2.4 Power spectrum of wind speed variation.

Fortunately, these variations, characterised by durations of typically three to five days, are slow in the context of the operation of large power systems. Apparently more difficult to deal with is the impact of short-term variations due to wind turbulence that are clear on the right-hand side of Figure 2.4. However, as will be shown later, the aggregation effects will reduce this problem considerably. Fortuitously it is the time scales at which there is least variation, the so called spectral gap between 10 minutes and an hour or two, that pose the greatest challenge to power system operation.

The essential characteristics of the long-term variations of wind speed can also be usefully described by a frequency or probability distribution. Figure 2.5 shows the frequency distribution for a year of 10 minute means recorded at Rutherford Appleton Laboratory, Oxfordshire, UK. Its shape is typical of wind speeds across most of the world's

Figure 2.5 An example wind speed frequency distribution.

windier regions, with the modal value (the peak) located below the mean wind speed and a long tail reflecting the fact that most sites experience occasional very high winds associated with passing storms. A convenient mathematical distribution function that has been found to fit well with data, is the **Weibull** probability density function (Eq. (2.3)), expressed in terms of two parameters, k, a shape factor, and C, a scale factor that is closely related to the long-term mean. These parameters are determined on the basis of a best fit to the wind speed data. A number of mathematical approaches of differing complexity are available to perform this fitting [7, 8].

$$p(U) = (k/C)(U/C)^{k-1} \exp[-(U/C)^k] \tag{2.3}$$

In practice, it is often more convenient to work with the cumulative distribution function representing the probability that the wind speed U is less than some value V (Eq. (2.4)); this is derived by integration of Eq. (2.3) between the limits zero and V.

$$Q(U < V) = 1 - \exp[-(V/C)^k] \tag{2.4}$$

The formula can be used to calculate the probability of the wind speed falling in a given range. For any two values of wind speed, U_1 and U_2, the probability of the wind speed being between U_1 and U_2 is simply $Q(U_2) - Q(U_1)$. This is used when calculating the expected yield from a wind turbine or wind farm at a given site.

It is also important to be able to calculate *extreme winds* at a given site. This also depends on use of a probability distribution, this time as given by Eq. (2.5) where $\hat{U}$ denotes an extreme wind or *gust* typically defined as an average over 3 s.

$$P(\le \hat{U}) = e^{-e^{-\alpha(\hat{U}-U_m)}} \tag{2.5}$$

$$R = \frac{1}{1 - P(\le \hat{U})} \tag{2.6}$$

where $1/\alpha$ is a measure of the spread of the distribution and U_m is the mode of the distribution. It is more usual to describe extreme values in terms of the return period R, defined in years by Eq. (2.6).

Note that the probability that an n-year wind speed will occur at least once in n years is not 1. In fact, the probability of exceeding $\hat{U}$ at least once in n years is $1 - [P(\le \hat{U})]^n$. If a wind turbine is intended to last N years the risk r, of exceeding its maximum design wind speed during that period should be suitably small, and is given by:

$$r = 1 - (1 - 1/R)^N \tag{2.7}$$

2.4.3 Wind Turbines

The power in the wind than can be extracted by a wind turbine is proportional to the cube of the wind speed and is given in watts by:

$$P = \frac{1}{2}\rho A U^3 C_p \tag{2.8}$$

where ρ is the air density, A is the rotor area, U is the wind speed and C_p is the power coefficient that represents the aerodynamic efficiency of the rotor. The variability in power output from one wind turbine would therefore be expected to substantially exaggerate the variability shown in the time histories of Figure 2.3.

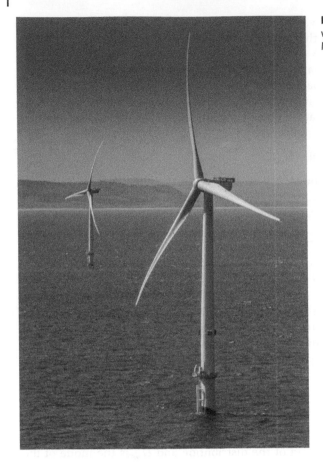

Wind turbines are designed to generate their rated or nameplate output at a *rated wind speed, U_r*. For wind speeds below a *cut-in wind speed, U_{ci}*, the wind turbine is not operational as the developed aerodynamic torque is not sufficient to overcome the frictional losses of the drive train and generate a useful power. For wind speeds above the rated wind speed, the power is controlled aerodynamically to maintain the output at the rated value until some limiting wind speed value is reached, known as the *cut-out wind speed, U_{co}*, at which point the turbine is shut down. The relationship between power and wind speed is known as a *power curve*. A power curve for the 8 MW Vestas wind turbine illustrated in Figure 2.6 is shown in Figure 2.7. This is the largest wind turbine commercially available at the time of writing. For this machine $U_{ci} = 4$ m/s, $U_r = 13$ m/s, and $U_{co} = 25$ m/s, values that are typical of large modern turbines.

This power characteristic combined with temporal variations in wind speed produces time varying electricity generation. Once the long-term wind speed variations have been expressed in terms of a probability distribution as in Eq. (2.3), this can be combined with the power curve to calculate the average output of the turbine or wind farm according to Eq. (2.9), where $P(U)$ is the power output from the wind turbine at wind speed, U. In practice, this calculation is usually undertaken as a summation and it can be more convenient to use the cumulative distribution of Eq. (2.4); details of this procedure can

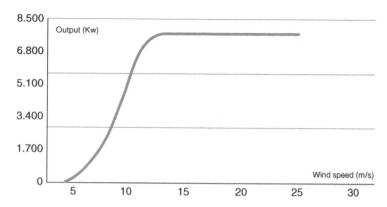

Figure 2.7 Power curve for Vestas V164 8.0 MW turbine. Source: Courtesy of MHI Vestas Offshore Wind.

be found in Reference [9]. Average outputs lie typically in the range of 0.25–0.45 of rated output depending on the mean wind speed at the site in question. These so-called *load factors* or *capacity factors* are much lower than would be expected for conventional generators and are discussed in some detail in Chapter 3.

$$\overline{P} = \int_{u=0}^{\infty} p(U) P(U) \, dU \tag{2.9}$$

The aerodynamic manner in which the wind turbine rotor extracts energy from the wind is described in a number of text books, for example References [10–12]. There is a well-defined upper limit to the aerodynamic efficiency C_p of a rotor; this is known as the Betz limit and is approximately 0.59; it reflects the fact that the air is not forced to flow through the rotor (e.g. as in a ducted turbine) but can flow around it instead. Conventionally, C_p is plotted as a function of the *tip-speed ratio*, λ (defined by $\lambda = \Omega R/U$ where Ω is the angular velocity of the rotor, R the radius and U the incident wind speed). λ is defined in this way to provide a generalised representation of the wind turbine rotor performance that is applicable to all combinations of incident wind speed and rotor rotational speed.

A typical C_p-λ characteristic is shown in Figure 2.8. It is apparent that to operate at peak efficiency (for large modern wind turbines C_p is normally in the range 0.4–0.5), the tip-speed ratio must be held constant for maximum output and this requires the rotor speed to be controlled in proportion to the wind speed. This is one reason why most large modern wind turbines are designed to operate at variable speed (see Chapter 4 for a more complete discussion). This is attractive from an integration perspective as the rotor has inertia available to absorb or release energy when accelerating or decelerating, respectively, thus smoothing short-term variations in wind speed. Consequently, its electrical power output varies less and can be more easily accommodated by the electrical system.

The C_p-λ of Figure 2.8 assumes a fixed blade configuration. If the blade orientation is changed, then the efficiency will change too. In fact, if the blades are *pitched* or *feathered*, that is, rotated about their axis so as to reduce the angle between the blade chord and the resultant incident wind, then the lift forces on the blade that produce the torque will reduce. This will result in a C_p-λ curve that sits below the one shown in Figure 2.8. The more the blades are pitched away from the optimal position, the lower the output at any

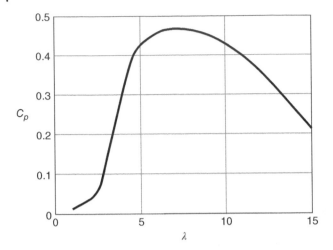

Figure 2.8 *Cp*-lambda curve for a three-bladed rotor.

given combination of wind and rotor speed. Wind turbines that can control their output by pitching the blades in this manner are *pitch controlled*. An alternative is to allow the process of stall to limit the wind turbine output. Such *stall regulated* machines have a fixed blade pitch and run at a nominally fixed rotor speed so that wind speeds above rated result in stalling of the blades, thus limiting power output around the rated value. Stall regulation was widely used for the smaller turbines (50–500 kW); but this passive approach to power control lacks the potential for operators to adjust the turbine output in response to external conditions. In part, as a consequence of this practically all MW size wind turbines are now pitch controlled in some way. An approach known as *active stall control* combines pitch and stall regulation. Dynamic power control is by stall but the blades are slowly adjusted to make sure stall occurs at the correct power level.

The mechanical shaft power created by the wind turbine rotor is converted to electricity by an electrical generator with a conversion efficiency that may reach 98% for large generators. Variable speed operation requires a frequency conversion through a power electronic converter, a process that reduces somewhat the overall efficiency. The generator and frequency conversion aspects are discussed in Chapter 3.

2.4.4 Power Variability

The short, medium and long-term variations in wind speed shown in Figure 2.3 affect power system operation in different ways. This is sufficiently important, given the significant and growing proportion of wind capacity in a number of power systems, to merit more detailed discussion.

2.4.4.1 Variability from Second to Second
At low or medium wind speeds the electrical output from a single wind turbine could vary substantially as a result of the turbulence in the wind. In Chapter 5, it will be shown that this may have a detrimental effect on the power system. However, when wind turbines are clustered in wind farms, there is physical spacing between them and the turbulence seen by each wind turbine is different and to a great extent uncorrelated.

The electrical output from wind farms therefore exhibits substantially lower relative variability than that from a single wind turbine. At the planning stage, appropriate analytical studies are carried out to ensure that the variability expected from a wind farm at a particular site will not adversely affect the power system.

2.4.4.2 Variability from Minute to Minute

Figures 2.3 and 2.4 indicate that the character of wind is such that if the second to second turbulence is removed, the average wind speed from 10-minute period to 10-minute period remains effectively constant. Chapter 3 will show that this 'persistence' nature of such averaged wind speeds is particularly important in integrating wind generated electricity in power networks. In practice, the output of turbines can be regarded as uncorrelated on the time scale of minutes and as with the faster variations the effect of aggregation is to smooth out variations at these higher frequencies.

2.4.4.3 Variability from Hour to Hour and from Day-to-Day

Figure 2.9 shows the actual records of wind speed at 13 geographically dispersed wind sites in the UK. As expected from Figure 2.4 there is substantial variability at each location over the 72 hour recorded period. Variability now becomes significant and effective integration of wind power in an electrical power network would have been problematic if the total resource were to fluctuate in sympathy to the fluctuations at one site. Fortunately, this is not the case due to the geographic diversity of the sites. The benefits of geographic diversity are clearly illustrated by the bold line in Figure 2.9, which represents the average wind speed from all sites. This average is substantially smoother (i.e. exhibits much reduced variability) than the wind at the individual wind farm sites.

Figure 2.9 UK site-specific and average hourly wind speed over 72 hours, with 50 or more records per hour. Source: Reproduced from Reference [25] with permission of the Environmental Change Unit, Oxford University.

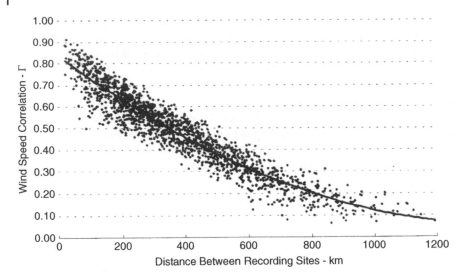

Figure 2.10 UK onshore wind power correlation by distance between sites based on UK long-term averages. Source: Reproduced from Reference [25] with permission of the Environmental Change Unit, Oxford University.

With large-scale exploitation of the wind resource, wind farms are installed inevitably across a variety of geographically dispersed sites. As suggested previously, this has a major beneficial effect in terms of overall variability. Just as the output from a wind farm has less short-term variability than a single wind turbine due to their dispersion across the site, so the aggregate output from several geographically dispersed wind farms has reduced longer-term variability than the output from a single wind farm. This reflects the fact that distant localities experience variations in wind due to shifting weather patterns that are time shifted in relation to one another and also to an extent distinct. Figure 2.10 presents the correlation between pairs of onshore wind sites in the UK as a function of the distance between the sites, and demonstrates that sites very far apart exhibit low cross-correlation. The data from Reference [13] were recorded over a period of 15–20 years.

2.4.4.4 Seasonal Variability
Seasonal and monthly average wind speeds vary significantly over most of the world. Figure 2.11 shows the seasonal changes of monthly averaged wind speeds for Billings, Montana in the USA. The trend of higher wind speeds during winter compared to summer is typical of the Northern hemisphere. The figure also indicates that there is variability from year to year.

2.4.5 Offshore Wind

Since 2000, offshore wind energy capacity grew significantly; so far, most of this is in Europe. There is a number of drivers for this: many countries have used up the best onshore sites, planning permission is often difficult to obtain onshore and the offshore wind resource is generally higher and the wind steadier (i.e. having less long-term time variation), also the turbulence levels are generally much lower. In addition, handling

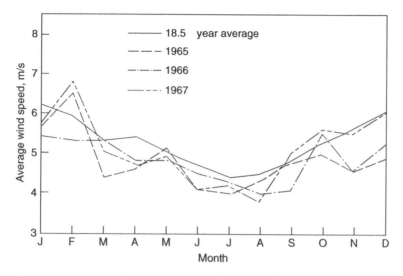

Figure 2.11 Seasonal changes of monthly average wind speeds. Source: Reproduced from Reference [12] with permission of John Wiley & Sons Inc.

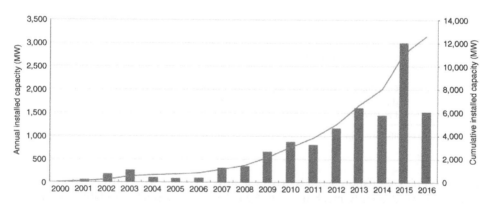

Figure 2.12 European cumulative and annual offshore installations. Source: Courtesy of WindEurope [14].

large blades and towers can be easier offshore than for transport by road. Figure 2.12 shows the year by year installation of European offshore wind capacity, taken from WindEurope, [14]. The total cumulative installed capacity exceeded 12 GW in 2016, and a sizeable amount is now under construction, especially in the UK.

Offshore wind is presently more expensive than onshore, primarily due to much higher costs for installation, and operations and maintenance. Although cost will vary significantly with distance from shore and local conditions such as water depth, sub-sea conditions and so on, a very rough guide used to be that offshore wind energy is roughly twice the cost at the point of connection to the power system although this is increasingly inaccurate as offshore wind costs fall. Table 2.1, taken from [15], shows an estimate of how the cost of offshore wind energy breaks down amongst the key categories.

Table 2.1 Overview of the cost of energy from offshore wind.

Sub-area	Descriptions	Percentage COE
Turbines	• Current turbines are less than or equal to 5–6 MW, with 3 blades on a horizontal axis, and most designs use gearboxes to drive the generator • Turbines are installed in arrays to create large wind farms	28%
Foundation	• <30 m foundations are usually simple steel tubes, e.g. monopoles, although larger turbines may use more sophisticated foundations • Foundations suited for 30–60 m water depth are more sophisticated than monopiles and are fixed to the sea floor, e.g. concrete gravity bases, tripods or jackets • Floating platforms could potentially be used in 60–100 m depths – e.g. tension leg platforms; various spar buoy concepts are being development outside of the UK for depths ≫100 m	19%
Collection & Transmission	• Currently high voltage AC (HVAC) cables are used to link turbines to an offshore substation, with power clean-up at each turbine • HVAC cables are also used to transmit power to the onshore substation as current wind farms are relatively close to shore within (60–80km)	13%
Installation	• Currently oil and gas vessels that jack-up from the seabed to install the foundation and turbine. Dynamic positioning (DP2) vessels have also been used to a certain extent, but this is not yet the norm	22%
O&M	• Currently access technologies involve helicopter transfers and direct boat access from shore that works best in calm seas • Limited remote condition monitoring	18%

Source: Courtesy of Offshore Wind [15]. But note that wind turbines are now larger and increasingly direct drive.

The economics of offshore wind is driving a move to larger wind turbines than would be used onshore. Presently, machines of 6–8 MW are commercially available, and larger machines are under development; in the future, turbines up to 15 MW could become available. Associated with this scale-up, different drive train arrangements are being used. Whereas MW turbines for onshore use still often use the Double Fed Induction Generator (see Section 4.6.2), offshore wind turbines are more commonly direct drive or hybrid configurations combining a low speed multi-pole permanent magnet generator (PM) with a low step up ratio gearbox. Table 2.2 compares the characteristics of a range of offshore prototypes that have been developed in recent years; not all of these reached commercial production, and in recent years some companies have been absorbed into others as indicated. The company Bard no longer exists. Presently, the offshore wind market is dominated by Siemens and Vestas.

Because the cost of operations and maintenance is much higher for offshore wind than onshore, this is an area deserving of more attention. The O&M figure of 18% in Table 2.1 is for present offshore installations. Future sites such those within the UK's Round 3 will be much more challenging, being both further from the land and also subject more extreme sea conditions often with mean annual significant wave heights of

Table 2.2 Characteristics of large turbines developed for offshore.

Manufacturer	Model	Date of full-scale prototype test	Rated power (MW)	Generator type	Gearbox stages
Repower	5 M	2004[a)]	5.0	DFIG	3
Senvion (prev. Repower)	6 M	2009[b)]	6.15	DFIG	3
Bard	5 M	2008[a)]	5.0	DFIG	3
Bard	Bard 6.5	2011	6.5	PM	3
Siemens	SWT 6–120	2011[b)]	6.0	PM	0
Sinovel	SL 6000	2011[b)]	6.0	DFIG	3
GE Energy	4.1–133	2011[b)]	4.1	PM	0
Guodian	UP 6000–136	2012[b)]	6.0	DFIG	3
Alstom (now GE)	Haliade 150	2013[b)]	6.0	PM	0
Gamesa (now Siemens)	G11X-5.0	2013[a)]	5.0	PM	2
Vestas	V164–7.0/8.0	2014[a)]	7.0/8.0	PM	2
Gamesa (now Siemens)	G14X	2014	7.0	PM	2
Adwen (now Gamesa)	AD 8–180	2017	8.0	PM	2
2BEnergy	2B6	2016	6.0	DFIG	3

0 in the last column indicates direct drive
a)   indicates commercial deployment,
b)   present commercial availability

over 2 m. A key issue affecting offshore O&M is restrictions on access. For more remote sites it may well prove impossible to access turbines for months at a time due to sea conditions. Even when personnel access is possible, say by use of helicopters, major repairs cannot be completed in high seas and winds due to the need for heavy lifting. Figure 2.13, taken from [16], shows how the expected delay time to complete a repair, which is caused by the need to wait for a suitable weather window, depends critically on both the time taken to effect the repair and also the significant wave height that the repair vessel can operate in. This example was calculated for the UK wind farm at Barrow. This was one of the Round 1 wind farms located near to shore and subject only to modest seas. Even for this relatively benign site many days of turbine operation can be lost waiting to undertake the more major repairs like generator or gearbox replacement. Future sites are expected to be worse, and it is anticipated that months of operation may be lost, resulting in low turbine availability. High onshore availability (typically 97–98%) depends to a high degree on a rapid response to faults, but this will not be easy to achieve offshore. It is thus essential that steps be taken to improve offshore availability through a combination of more reliable design, design for quick and easy maintenance and repair, and improved access and repair methods. Preventative maintenance based on effective condition monitoring of the turbines is expected to become important.

Worked Example 2.1
A site has a surface roughness of 0.1 m. The average wind speed measured at 10 m height over a year was 6.4 m/s. Calculate the expected annual average wind speed at a turbine hub height of 80 m, and estimate the turbulence intensity.

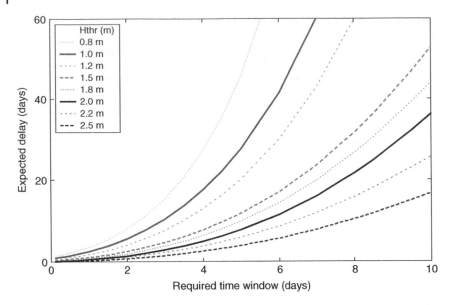

Figure 2.13 Dependence of expected repair delay on repair time and vessel operational limit.

Model Answer
From Eq. (2.1):

$$U(80) = U(10) \left[\frac{ln(80/0.1)}{ln(10/0.1)} \right] = 6.4 * 6.68/4.605 = 9.3 m/s$$

The turbulence intensity can be estimated using Eq. (2.2): $I(z) \approx \frac{1}{ln(80/0.1)} = 0.15$ or 15%. It should be noted that, unlike the extrapolation of wind speed to a different height, the estimate of turbulence intensity is not based on any wind measurements, and relies totally on the value of the roughness length. For this reason it should be considered as a rough estimate.

Worked Example 2.2
A 5 MW wind turbine has a cut-in wind speed of 4 m/s, a rated wind speed of 12 m/s and a cut-out wind speed of 25 m/s. It is located on a site with a Weibull shape parameter of 2.1, and a scale parameter of 8.8. If the average power of the turbine when operating below rated power is 2 MW, calculate the capacity factor if the turbine has an availability of 97%.

Model Answer
Equation (2.9) can be broken up into two parts that make a positive contribution to the mean power corresponding to generation below and above rated wind speed:

$$\overline{P} = \int_{u=0}^{\infty} p(U) \, P(U) \, dU = \int_{4}^{12} p(U)P(U)dU + \int_{12}^{25} p(U)P(U)dU$$

Since the average power output below rated wind speed is known, it can be treated as a constant and taken outside of the integral. The same applies to above rated wind speed, resulting in:

$$\overline{P} = 2 \int_4^{12} p(U)dU + 5 \int_{12}^{25} p(U)dU$$

The easiest way to estimate the two integrals is to use the cumulative Weibull probability distribution given by Eq. (2.3). In general, by the definition of the cumulative distribution:

$$\int_a^b p(U)dU = Q(b) - Q(a)$$

Application of this with Eq. (2.3) gives:

$$\overline{P} = 2 \int_4^{12} p(U)dU + 5 \int_{12}^{25} p(U)dU =$$
$$2 * (1 - \exp[-[12/8.8]^{2.1}] - 1 + \exp[-[4/8.8]^{2.1}]) + 5 * (1 - \exp[-[25/8.8]^{2.1}]$$
$$- 1 + \exp[-[12/8.8]^{2.1}]) =$$
$$2 * (-0.0147 + 0.826) + 5 * (-1.28 * 10^{-4} + 0.147)$$
$$= 1.62 + 0.735 = 2.355 MW$$

If the availability is 97%, then the capacity factor = 2.355*0.97/5 = 0.456

2.5 PV and Solar Thermal Electricity

2.5.1 The Resource

The average intensity of light outside the atmosphere (known as the solar constant) is near to $1353 \, W/m^2$. Attenuation by the atmosphere results in peak intensity at sea level of around $1 \, kW/m^2$, giving a 24-hour annual average of $0.2 \, kW/m^2$ averaged over the planet's surface. As this overall energy density is relatively small, large areas will be needed for a significant energy production: for example, in order to produce a gigawatt of power, an area of nearly $5 \, km^2$ would be needed (assuming a conversion efficiency of 20%). However, in countries such as the UK (with low irradiance[7] and high population density) the existing and appropriate roof space is sufficient in principle to generate enough electricity to cover a significant proportion of electricity consumption.

Common sense tells us that irradiation varies regionally, with the changing seasons, and hourly with the daily variation of the sun's elevation. Many locations do not experience unbroken sunshine. Cloud cover can significantly reduce the net radiation and cause relatively fast variations in intensity, in some cases significant variations from minute to minute or even over seconds.

Seasonal variation on the earth's surface is illustrated here by comparing two different sites. The magnitude will obviously vary from site to site but the basic principles are the same. Sites of similar latitudes should, in principle, have a similar solar resource.

The earth's tilt angle leads to a variation of the seasonal irradiance. Figure 2.14 illustrates the differences in average monthly irradiation for a site close to the equator,

---

7 Sometimes the terms solar radiance, radiation intensity, insolation (in older text books) can be found although irradiance is the most widely accepted. The SI units for irradiance are W/m^2.

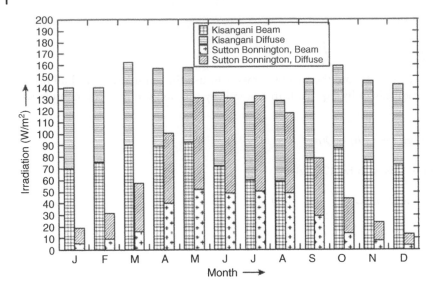

Figure 2.14 Comparison of Irradiation at Kisangani (Congo) and Sutton Bonnington (UK).

Kisangani (Congo, latitude −0.31°) and a site at higher latitude, Sutton Bonnington (UK, latitude 52.5°). As expected Kisangani receives twice as much irradiation in the course of the year. While Sutton Bonnington experiences a strong seasonal variation of ±73%, the irradiation in Kisangani varies only by ±13% around the yearly average.

To calculate the total radiation falling on a fixed surface of a given orientation at a given site (that could be PV module or a solar thermal collector) it is necessary to separately calculate the direct (sometimes call beam) and diffuse components. For calculation of the direct component on the surface, the position of the sun relative to the surface, as given by the incidence angle, must first be calculated. The zenith angle θ_z (the angle between the sun position and the vertical) is given by:

$$\cos \theta_Z = \sin \delta \cdot \sin \Phi + \cos \delta \cdot \cos \Phi \cdot \cos \omega$$

where δ is the declination angle that reflects the tilt of the earth's rotation with respect to its orbit around the sun, and varies with the time of year, ω is the hour angle, which is the angular displacement of the sun east or west of the local meridian due to rotation of the earth around its axis and Φ is the latitude.

In the Northern hemisphere, the optimal surface orientation is due south and inclined at an angle to the horizontal equal to the local latitude. In this case, the incidence angle θ is given by:

$$\cos \theta = \cos \delta \cos \omega$$

If the direct (beam) and diffuse radiation intensity on a horizontal surface are, G_b and G_d, respectively, and if reflection from the ground is ignored, then the radiation on the collector surface G_t can be calculated from:

$$G_t = G_b(\cos \theta / \cos \theta_Z) + G_d(1 + \cos \beta)/2$$

where β is the tilt of the surface. Full details of these calculations can be found in text books such as [17].

The annual average irradiation peaks in the north-east of Africa at around 300 W/m^2. Other favoured regions can be found all over Africa, the north of South America and the south of North America, and south-west Asia. Australia experiences solar irradiation levels above 250 W/m^2 across nearly half the continent. The irradiation reduces, generally speaking, the further one moves away from the equator where it can be well below 100 W/m^2 on a 24-hour average. Obviously, these values are averaged over the year and will vary significantly with the seasons. Most northern European, north American and north Asian countries are within this latter region and present and future systems are likely to be rooftop installations in the kilowatt scale connected to the local 230 or 115 V network.

2.5.2 The Technology

There are two main technologies for the conversion of sunlight into electricity. *Photovoltaics* (PV) depends on the use of semiconductor devices for the direct conversion of the solar radiation into electrical energy. Efficiencies of the typical commercial crystalline PV cells are in the range of 12–18%, although experimental cells have been constructed capable of over 30%. In contrast, *solar thermal systems* depend on intermediate conversion of solar energy into thermal energy in the form of steam that in turn is used to drive a turbo-generator. To obtain high temperatures, thermal systems invariably use concentrators either in the form of parabolic troughs or thermal towers. At present, generation of electricity by either technology is somewhat more expensive than traditional means. Due to the considerable potential of cost reductions in PV systems it is believed that in the future, perhaps in a decade or so from the time of writing, PV systems will be providing a sizeable proportion of the renewable energy contribution.

Some countries, notably Japan and Germany, have created substantial home based markets for PV, and the last couple of years have seen a rapid growth in multi-megawatt installations in Europe. PV costs have fallen dramatically in recent years and are now increasingly competative with other forms of generation, even in locations with a modest solar resource. In sunnier places such as California, Australia, North Africa and the Mediterranean, where the peak electricity demand occurs in summer due to tourism and air-conditioning loads, large-scale multi-megawatt PV plants or solar thermal plants are even more attractive. PV technology remains far from fully developed and new and cheaper cells can be expected in the future supporting significant contributions to electricity supply.

The characteristics of PV and solar thermal plant are rather different and are outlined next.

2.5.3 Photovoltaic Systems

At the heart of a PV system is the PV module. Detailed descriptions of the different PV technologies and the basics of solar cell operation can be found in a wide range of text books, for example References [18, 19]. PV modules produce output determined mainly by the level of incident radiation. They are characterised for given external conditions, by an *I-V* curve of the type shown in Figure 2.15. PV cells generally behave like a current source (dependent on the radiation) and a diode in parallel. The useful (load) current I_L is thus given by the difference between the generated current (photo-current) I_G and the junction (dark) current I_J.

$$I_L = I_G - I_J$$

$$= I_G - I_O \left[\exp \left\{ \frac{q}{AkT}(V + I_L \cdot R_S) \right\} - 1 \right]$$

where:

I_O	is the dark reverse saturation current of the diode, proportional to $\exp(-E_g/kT)$
q	is the charge on an electron (1.6×10^{-19} C)
K	is the Boltzmann Constant (1.38×10^{-23} J.K^{-1})
T	is the absolute temperature of the cell (K)
A	is the constant between 1 and 2 (varies with type of cell)
V	is the terminal voltage (V)

As the operating voltage increases the diode current increases and the output current decreases as shown in Figure 2.15. The power, $I \times V$, depends on the operating point and is maximised for operation near to the knee of the *I-V* characteristic, known as the Maximum Power Point (MPP). Power electronics is used to convert the dc output of the PV modules to ac for injection into the network (more about this in Chapter 3). The quality of a cell can be judged by the squareness of the *I-V* characteristic. This is quantified in terms of the ratio of the voltage at open circuit, i.e. where the *I-V* curve meets the voltage axis times the closed-circuit current where the *I-V* curve meets the current axis, divided by the power at the MPP. This ratio is known as the fill factor.

As shown in Figure 2.16, radiation and temperature affect the shape of the *I-V* curve and the voltage at which the MPP occurs. The power electronic converter is also used to control the operating voltage as near as possible to the MPP, and tracks this as it changes with radiation level and to a lesser extent with module temperature. Commonly, both these power electronic functions are combined in the inverters used for grid connection. PV modules have negligible energy storage capability and, as far as the power system

Figure 2.15 An example *I-V* curve.

Figure 2.16 Impact of radiation and temperature on the IV characteristic.

Figure 2.17 Time variation of radiation through a summer's day in Loughborough, UK.

is concerned, their output varies instantaneously with changes in radiation. Hence, a passing cloud, as occurs in Figure 2.17 just before 18:00, will result in the sudden collapse in output from a PV system. As with wind turbines, a geographical spread of PV systems will substantially mitigate such short-term effects.

It has been speculated that as power systems become increasingly interconnected, as is currently happening in Europe, it may eventually be possible to smooth out the diurnal variations in PV output[8]. In such a vision, significant intercontinental power flows might occur, for example, power generated from massive PV arrays in the Sahara during hours of daylight could be transmitted to areas of the planet then in darkness. Of course, this would involve power systems spanning substantial proportions of the planet; something that may never turn out to be cost effective.

---

8 The benefits of interconnection are central to the operation of power systems and will be returned to again and again throughout this book.

Figure 2.18 A 50 MW parabolic trough plant with thermal storage at Andasol, Spain. Source: Reproduced from Concentrating Solar Power – from research to implementation. © European Communities, 2007.

2.5.4 Solar Thermal Electric Systems

Solar thermal electricity generation systems most commonly use solar concentrators to produce high temperatures that can drive heat engines with acceptable conversion efficiency. Two main types of generator suitable for large-scale generation have so far been demonstrated, both requiring direct or beam radiation. Climates with cloudy or overcast conditions, where most of the radiation is diffuse, are thus unsuited to this technology. The two technologies are:

- Large collections (or solar farms) of parabolic trough reflectors, Figure 2.18, focus solar radiation onto a line receiver containing a piped heat transfer medium. The medium can be a thermal oil, capable of withstanding high temperatures, or alternatively pressurised water. This is collected and passed through a heat exchanger/boiler where steam is raised for supply to the turbines. Operational temperatures vary between 350 and 400 °C and system sizes up to 80 MW have been built. To increase steam turbine operating temperature, and thus efficiency, steam from the solar system can be further heated using conventional fuels.
- In solar power towers a central receiver is mounted on top of a tower that is surrounded by a field of concentrating mirrors (heliostats) that track the sun. The heliostats reflect and concentrate the light onto a receiver where the energy is absorbed by the heat transfer medium, which could be water, a molten salt or any other suitable high temperature heat transfer medium. Temperatures as high as 1000 °C can be achieved. System sizes up to 200 MW have been considered but to date the largest systems constructed are considerably less. Figure 2.19 shows one of the larger systems currently operating, the 11 MW PS10 solar power tower plant at Sanlúcar la Mayor, Spain. Solar power towers are expected to be more economic than solar farms for larger-sized plant, say over 100 MW.

Since both these technologies depend on direct sunlight for efficient operation they function for only part of the time, and then not always at maximum output. However,

Figure 2.19 Aerial view of 11 MW PS10 solar power plant at Sanlúcar la Mayor, Spain.
Source: Reproduced from Concentrating Solar Tower – from research to implementation.
© European Communities, 2007.

because they both involve a thermal intermediary stage, they can be combined with fossil fuel combustion (a hybrid) and depending on design details it may be possible to incorporate thermal storage. The primary advantage of these adaptations is the capability to supply power when the sun doesn't shine, in other words to be dispatchable.

A smaller scale approach, suitable for units in the range 10–50 kW, is possible using parabolic concentrators. These also work at high temperatures of 600–1000 °C and might be suitable for distributed electricity generation.

Thermal-to-electric efficiency is in the range of 20–40%, depending on the design, with the resulting overall solar-to-electricity conversion efficiency in the range 13–25%.

Table 2.3 (adapted from material available from the website given in Reference [20], when it was accessed in January 2006) compares the performance of the parabolic trough and power tower systems.

A good up to date review of solar thermal electric systems can be found in Reference [21]. Although considerable cost reductions are anticipated for the technology as it matures, up to 60% for the cost of energy, it will be hard for the technology to compete with photovoltaics. A key advantage though is the lower cost manner of introducing storage, through low cost thermal storage as described before. This advantage will become of increasing importance as renewable energy penetrations into power systems increase.

Table 2.3 Performance of parabolic trough and power tower systems.

	Parabolic trough	Power tower
Size	30–320 MW	10–200 MW
Operating temperature (°C/°F)	390/734	565/1049
Annual capacity factor	23–50[a]%	20–77[9]%
Peak efficiency	20(demonstrated)	23(predicted)
Net annual efficiency	11(demonstrated)–16%	7(demonstrated)–20%
Status	Prototype	Demonstration
Storage available	Limited	Yes
Hybrid designs	Yes	Yes

a) The top end capacity factors are for systems including substantial thermal storage.
Source: http://www.eere.energy.gov/consumerinfo/pdfs/solar_overview.pdf.

2.6 Tidal Power

2.6.1 The Resource

The moon and sun's gravitational fields cause the natural rise and fall of coastal tidal waters. Since the moon is closer to earth, albeit far less massive, it has a dominant effect on tides. The moon is 2.2 times more influential than the sun, thus we could consider tidal energy to be mostly a form of lunar energy!

The earth rotates on its axis once every 24 hours. In the earth's frame of reference, the sun orbits the earth once every 24 hours. The moon orbits the earth once every 29 days approximately. In the earth's frame of reference, the moon appears to orbit the earth once every 24 hours and 50 minutes. This difference in periods between the apparent orbits of the sun and moon leads to phase changes with larger *spring tides* during in phase behaviour and smaller *neap tides* when the sun and moon are out of phase. Spring tides can be twice as large as neap tides.

An early study commissioned by the EU estimated that the technically feasible energy resources from tidal barrages across the EU could be as much as 105 TWh/year (from 64 GW of installed capacity). This resource is unevenly distributed across Europe with the UK (47.7%) and France (42.1%) sharing the bulk of the resource and Eire (7.6%) accounting for most of the rest.

It has been estimated that exploitation of all practicable estuaries in the UK could lead to electricity generation of up to around 20% of demand in England and Wales. Despite this potential, very little has been done to date to exploit the resource.

2.6.2 Tidal Enhancement

Over much of the surface of the oceans, the *tidal range* (the vertical rise and fall) is rather small, less than 1 m, but in certain places, there is an enhancement of the range. Enhancement may be due to:

2.6.2.1 Funnelling
The tide is gradually constrained from the sides and so increases in height – or the reverse, later in the cycle.

2.6.2.2 Resonance

The estuary has a resonant period equivalent to the tidal period. The length and depth of the estuary are very important for resonance.

2.6.2.3 Coriolis Effect

The spinning of the earth leads to the Coriolis effect mentioned early with respect to wind. The tide is influenced by this and in some locations tends to increase in height at high tide and be drawn away from the coast at low tide, with the net effect of enhancing the tidal range.

These enhancements are highly predictable, and so permit us to largely determine the output from a tidal scheme for years ahead.

2.6.3 Tidal Barrages

These work by constructing a barrage across an estuary and allowing tidal waters to alternately fill the estuary through sluice gates and then to empty it through turbines that can generate energy. A barrage constructed across an estuary is equipped with a series of gated sluices and a bank of low-head axial turbines. Where it is necessary to maintain navigation to the upper part of the estuary, a ship-lock may be required.

Tidal barrages are a currently available technology, but very few exist worldwide. The best-known example is the 240 MW scheme at La Rance in France, and smaller installations have been made in Nova Scotia, Russia and China. The UK has a number of attractive sites due to its high tidal ranges, the largest potentials being on the Severn estuary (8600 MW capacity), and the Mersey (700 MW capacity). However, the scale of these installations and the calculated long payback periods make the required investments unlikely in the context of the privatised electricity supply industry. Furthermore, environmental considerations may present a barrier to large-scale developments.

2.6.4 Operational Strategies

Ebb generation is the simplest mode of operation for a tidal barrage scheme. The operating cycle consists of four steps:

- Sluicing on the flood tide, to fill the basin.
- Holding the impounded water until the receding tide creates a suitable head.
- Releasing the water from the basin to the sea via turbines, on the ebb tide, until the tide turns and rises to reduce the head to the minimum operating point.
- Holding until the tide rises sufficiently to repeat the first step.

Ebb generation with flood pumping is a modification of this mode that allows increased energy output. By using the turbines in reverse as pumps, the basin level and hence the generating head can be raised. The energy required for pumping must be imported but since the pumping is carried out against a small head at high tide and the same water is released later through the turbine at a greater head, this can produce a net energy gain with some limited ability to re-time output. The energy gain through pumping could be small but useful and typically in the range 3–13%.

Flood generation is the reverse of ebb generation and is rarely suggested alone, possibly because it offers little storage opportunity as the basin is often being filled by a river, which reduces the total energy capacity.

Two-way generation (ebb and flood) is possible with reversible turbines and is used at La Rance (together with flood pumping). The additional energy recovered may not justify the extra cost and complexity of the turbines.

Most schemes propose ebb only generation. The full resource could only be captured if the barrage of the scheme has a very large two-way generating flow capacity in order to achieve rapid filling and emptying of the basin. In practice, it is neither economic nor is there enough physical space to install enough turbines to come anywhere near full resource capture. Typically, only about one-third of the gross power can be harnessed, so after turbine and generator losses the mean electrical output will at best be about a quarter of the gross power.

The other main reason for preferring ebb generation is the nature of the amenity that results. In such a scheme, the water level in the basin never falls below the mean sea level, so the basin is available for recreational activity much like an inland reservoir. On the other hand, it is bad news for the bird population for whom exposed mudflats at low tide provide a major food resource. High initial capital cost and polarised opinions on the resulting environmental changes are the main reasons why tidal schemes seldom get beyond the feasibility study stage. Tidal energy barrages are expected to have very long lifetimes. Their design life could be about 120 years but with normal maintenance and replacement of turbine generators at 40-year intervals, their lifetime could effectively be unlimited.

2.6.4.1 Power Variability

One of the main disadvantages of tidal schemes is the pulsed nature of their electrical output. Figure 2.20 shows the expected electrical output from the 7 GW scheme proposed for the Severn estuary assuming ebb generation only.

Power is generated for 5–6 h during spring tides and for 3 h during neap tides. This pattern may, but is unlikely to match the pattern of demand in an integrated power system. On the positive side the power availability from such a scheme is highly predictable.

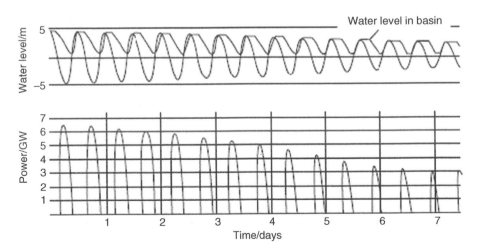

Figure 2.20 Water level and electrical power output of the proposed Severn Barrage over a spring-neap tide cycle. Source: Watt Committee on Energy (now disbanded). Reproduced with permission of Oxford University Press.

In a large integrated system such as the UK's, this predictability will ease the task of scheduling alternative generation to take up the demand when the output from the scheme falls to zero.

2.6.5 Tidal Current Schemes

The direct production of electricity from tidal streams is a relatively new concept. It relies on a different approach to conventional tidal barrage schemes, and is attractive in that it does not require such massive infrastructural investment.

Tidal stream technology extracts energy directly from the currents that flow in certain locations, driven by the rise and fall of tides in the vicinity. These currents usually have a low velocity (1 m/s) though this can be enhanced by the local topography. In particular, the velocity can be magnified greatly in straits between islands or between islands and the mainland. Tides can be predicted with very high accuracy; hence after measurements at a site, the energy available for conversion can be forecast with confidence.

Figure 2.21 shows a tidal cycle together with the associated stream current velocity. Tidal height varies approximately sinusoidally and as already mentioned, is more or less completely predictable. For any site, the tidal current velocity has four peaks and four troughs per day, as in the figure.

The technology to extract energy from the tidal current is conceptually simple: a turbine is placed in a suitable tidal flow, which turns the generator through a gearbox as in Figure 2.22. It is similar to a submerged wind turbine, except that the greater specific gravity of seawater results in much higher energy densities in tidal streams than is found in winds of the same velocity. However, the water velocities available in tidal streams (typically rated velocities of 2–3 m/s on good sites) are much lower than the air velocities used by wind turbines. Although power output is proportional to the cube of the velocity, tidal stream rotors generally produce significantly greater output than wind turbines of the same size because of the massively increased water density. Compared to

Figure 2.21 Tide height and current velocity. Source: Reproduced from Reference [25] with permission of the Environmental Change Unit, Oxford University.

Figure 2.22 Artist's impression of axial flow tidal current turbines Source: Reproduced with permission from Marine Current Turbines Limited.

wind, tidal flow velocities were initially expected to have little turbulence and thus vary in a smooth manner, reducing fatigue loads on the rotor and generating electricity with little short-term variation. Recent site assessment data, however, does not support this. Sub-sea surfaces in locations with high tidal flows have often been found to be scoured of sediment, with resulting high levels of turbulence.

Following at least a decade of technology development and prototype testing, tidal stream technology is now on the cusp of commercial deployment. The MeyGen project started the process of installing the first four of its 1.5 MW turbines in the Pentland Firth (located between the Scottish mainland and Orkney Island) in September 2016. When complete, the site will have a total of 269 such turbines. The site is exceptional with peak flow rates of up to 5 m/s due to the channelling of tides between the mainland and Orkney; most of the attractive tidal stream sites are enhanced by flow channelling, but few will reach such high flow velocities.

At different sites, the peaks and troughs occur at different times. Power being proportional to the cube of current velocity, this implies that maximum and minimum electricity generation is possible at different times. By combining generation from sites that are out of phase it is theoretically possible to smooth out the diurnal variability. The tidal stream resource is highly site specific, and there are a limited number of tidal stream sites around the world worth exploring. The accessible tidal stream resource for the most suitable sites in the UK (including the Channel Islands) is estimated to be approximately 36 TWh/year. The power variability from such schemes will be similar to that from barrages.

A second type of variability also affects tidal velocities. This is the spring/neap cycle, which occurs over 14 days and changes tide ranges (the difference in height between high and low tide) from a minimum to a maximum, then back to a minimum. Unlike diurnal variations, the spring/neap cycle affects all sites in the same way at the same time, so there is no scope for smoothing by combining generation from many sites.

It can be concluded that patterns of tidal stream power output differ significantly from electricity demand cycles; maximum potential output sometimes coincides with peak demand but at other times minimum output does. But at the likely levels of penetration from this resource, the magnitude of this variability is small compared to demand variations.

2.7 Wave Power

2.7.1 The Resource

The passage of wind over the surface of the sea results in the gradual transfer of energy into the water to produce waves, so wave energy too is an indirect form of solar energy. Wind power typically has densities in the range $1.2–1.8\,\mathrm{kWm^{-2}}$. Waves with a typical power density of $50\,\mathrm{kW}$ per metre of wave front or crest length are in effect a highly concentrated form of solar energy. The distance over which this process of wave generation occurs is called the 'fetch' and longer fetches produce larger, more powerful waves as do stronger winds and extended periods of wind. Sea waves are characterised by their wave height (H), period (T) and crest length. The power per metre of crest length is proportional to the period and to the square of the wave height.

Wave records indicate that the heights and periods of ocean waves vary continuously over time, which results in wave power varying almost continuously over time. However, the conversion of the energy of a wave face into electricity output tends to partly smooth this variability. Because of these irregularities the calculation of the power density requires an averaging process. The *significant wave height, Hs*, is approximately equal to the average of the highest one-third of the waves and the *zero-up-crossing period, Te*, is defined as the average time-counted over 10 crossings or more- between upward movements of the surface through the mean level. It can be shown that the average power in 1 m of wave crest is proportional to *Te* and to the square of *Hs*.

The world wave resource is not yet fully analysed but there is no doubt that it is sizeable. Table 2.4 gives an estimate of the scale of the technically achievable resource.

2.7.2 The Technology

The conversion of wave power into electricity requires a device that intercepts the waves and converts a proportion of their energy first into mechanical and then into electrical form. The conversion of wave energy into mechanical energy demands a central

Table 2.4 Technically achievable wave power resource.

GW	UK potential	World potential
Shoreline	0.03	1–50
Near shore	0.3–0.7	10–500
Offshore	7–10	200–5000

stable structure incorporating an active element that moves relative to it under the forces exerted by the waves, and can react against the central structure to produces forces and displacements that generate mechanical power.

Wave energy research and development has been taking place for over 25 years now, and significant progress has been made towards the development of viable technologies able to exploit the large energy potential of the world ocean wave climates. Ocean waves are often powerful, but with extremely low frequencies, of about 0.1 Hz (equivalent to 6 rpm or to periods of around 10 s), and the success in generating electricity demands that this frequency is raised to 500–1500 rpm.

The technology for wave energy conversion is still at an early stage compared to wind and photovoltaics and a variety of different approaches are currently under development. Because of the nature of the resource and for efficient conversion, the swept volume of the device must be of the order of several tens of cubic metres per metre of device width. The devices are therefore physically large and have to be designed to withstand without damage extreme waves that may occur very rarely. The transition of a concept from a model tested in a laboratory wave tank to a working prototype requires considerable expenditure and it is only very recently that serious funding from governments and industry has been forthcoming.

It is not the intention here to describe in detail the various concepts and associated hardware. The devices can be classified in a variety of ways depending on their intended location or their geometry and orientation. At least 12 different devices are being developed worldwide and it is still too early to guess which of them might be capable of providing energy competitively and reliably. A particular technology that seemed promising at one time was Pelamis; it was designed to ride through extreme seas without creating excessive loads on the structure, and a 750 kW commercial scale unit was developed. In 2004 it became the first wave power generator to operate grid connected, but it ultimately proved too expensive and the company, Pelamis Wave Power, went out of business in 2014. This highlights the considerable challenge of developing economically viable wave energy technology.

While the UK is fortunate in having a good wave climate, the political climate has not always favoured the technology. Attitudes, however, are changing, prompted by the need to address global climate change, the long-term resource security of fossil fuels and by the increasingly competitive economics of wave energy. UK R&D teams conducted much of the early work, but many other countries are now active in wave energy development. Commercial involvement is now significant and schemes or concepts from The Netherlands, Norway, Australia, Sweden, Denmark and the USA as well as the UK are now being developed. Nevertheless, the challenges of producing cost effective technology that can withstand the extremely challenging operational environment should not be underestimated.

2.7.3 Variability

The power output from a single wave power device will follow an exaggerated variation of the wave height changes because of the square relationship between power and wave height. Not unlike offshore wind farms, a substantial wave power installation will consist of an array of such devices the outputs from which, when added together, because of spatial effects, will be relatively smoother than that from one device. If this were extended to hundreds of such devices in geographically dispersed arrays, the overall output will be smoother still. The short-term variability of the output from wave power converters is unlikely to present integration problems assuming the output from arrays are connected to a grid voltage level capable of absorbing variable generated power without adverse network effects.

The wave resource, not unlike the wind resource on which it depends, also varies on a day-to-day and season-by-season basis; in general, wave conditions are more energetic in the winter than in the summer. For example, about half of the annual wave power at all UK sites occurs during the winter months of December, January and February, as shown in Figure 2.23. This exceptional match between availability and demand for the UK would also hold for many countries with coastal areas facing the Atlantic or Pacific in the Northern hemisphere. Figure 2.23 also depicts the seasonal variability of wind to illustrate, by comparison, the pronounced variability of wave power.

Wave power levels, are predictable to varying degrees over different time scales. At short forecast periods (a few hours), information on the wave conditions offshore coupled with the power available from wave devices in the previous hour can be used to estimate future output. For longer forecast periods, wave models provide estimates of wave power up to five days ahead based on predicted wind speeds. More recent wave forecasting research, for example [22], explores the production of probabilistic forecasts that can be more useful to operators.

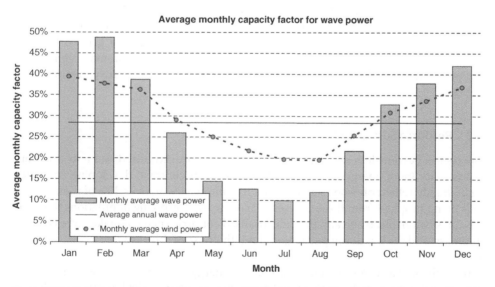

Figure 2.23 Average monthly wave power in the UK. Average monthly wind power capacity factor is shown for reference purposes. Source: Reproduced from Reference [25] with permission of the Environmental Change Unit, Oxford University.

2.8 Biomass

2.8.1 The Resource

Biomass differs considerably from the other renewable sources in that it takes the form of a fuel that can be stored and used for electricity generation when required, in the same way as fossil fuels. However, unlike fossil fuels, biomass is often limited by the energy density of the stored fuel. Therefore, it must be produced and consumed locally, as energy consumption associated with transportation over long distances might even exceed that of the fuel itself. This means that biomass power generating units are relatively small compared to conventional plant (relying on local supply chains for feedstock) and possess the characteristics of small embedded generating units.

There are three basic thermo-chemical conversion technologies that use solid biomass as a primary fuel for the production of electricity, namely direct combustion, gasification, and pyrolysis. In addition, the use of liquid biomass (such as sewage sludge) for the production of methane via anaerobic digestion is increasingly common.

Electricity production using solid biomass fuels is still a developing industry and as a consequence is not competitive on price with electricity from fossil fuels without some kind of government fiscal or policy support. However, it is competitive with nuclear power and possibly new-build clean coal power stations but not with modern gas fired power stations within the current regulatory and economic climate. With the correct support, as existed in the UK through the RO, co-firing of coal with biomass was commercially attractive. In the longer term, grid connected biomass generation (using the full range of possible technologies), may become competitive; the greatest potential is for small-scale embedded generation using gasification, pyrolysis or high-speed steam engine-based plant. In the short-term, small-scale (100–500 kWe) dedicated plants for use on farms or by rural industry has the greatest potential. In the medium term when increased demand for electricity could be causing the grid to become overloaded and unreliable, then larger (1–20 MWe) embedded biomass generation plant providing end-of-grid support may become an attractive alternative to reinforcing the grid.

2.8.2 Resource Sustainability

Biomass has become less important as countries have industrialised and now accounts for less than 3% of energy in the developed world. In contrast, developing countries remain highly reliant on wood and other natural biomass with over 30% of their energy needs being supplied from these sources. Of course, with growing populations this supply is not sustainable. Ironically, the industrialised countries need to make increased use of biomass, and the less developed regions limit their use of this resource to a sustainable level.

One of the key features of biomass is that the energy expended in growing it, that is, planting, watering, use of chemicals and pesticides to enhance yield, harvesting, drying and so on, is not negligible. Specifically, for ethanol production used as fuel for transport, the refineries themselves are fired by fossil fuels to ferment the crop and to purify ethanol from the product of fermentation. A US Department of Agriculture report found that the energy from corn bio-ethanol was a mere 8% in excess of the input production energy

Table 2.5 Energy ratio of several crops.

Crop	Ratio
Miscanthus	32.5
Willow	30.0
Hemp (straw)	8.5
Wheat (grain)	8.8
Oilseed rape	3.8

and a paper in *Science* [23] found that the energy ratio was only net-positive when the energy savings from 'co-products' for cattle feed were included. Efforts are now being made to produce bio-ethanol from cellulosic crops and not from fermentation. This promises to produce twice the amount of ethanol per hectare of crop.

The scene is a lot brighter if biomass is used to generate electricity, especially in combined heat and power (CHP) plants in small decentralised power stations. The benefits are compounded if the crops are grown organically, if possible, and used locally. The choice of crop is also vital in the effectiveness of CO_2 mitigation. Table 2.5 compares the ratio of energy out to energy in for a number of crops based on DTI report URN 01/797 [24]. Miscanthus and Willow are the preferred crops in European and other similar latitudes. Currently, the main use of miscanthus in the UK is in co-firing with coal in existing power stations

Large-scale biofuel production is not only energy intensive but it could have adverse environmental and social impacts. Such development requires substantial water resources with the result that water tables in areas of intense cultivation have been lowered to unacceptable levels. Expansion of biofuel crops could also speed up tropical deforestation with the associated lowering of CO_2 absorption and threat to extinction of thousands of species of animals and plants. If such crops are encouraged through subsidies, food shortages may occur if land previously used to produce food is lost. It may be concluded that the biomass path, unless used wisely, may cause serious environmental impacts.

2.9 Summary of Power Generation Characteristics

This chapter has reviewed the main forms of electricity generation. For the renewable energy forms that are the main topic of this book, the resource characteristics have been presented and discussed in some detail since they impact significantly on the manner in which theses sources can be integrated into the power system. We conclude by summarising the key characteristics of the different forms of generation in Table 2.6.

In this table, typical unit sizes for the different forms of electricity generation are referred to. These unit sizes are in all cases the nameplate rating of the prime mover/generator package. Power stations may contain one or several such units. The table also describes the nature of the energy resource in terms of availability. Traditional generators supplied from coal, oil, gas or nuclear energy score highly here

Table 2.6 Generator characteristics by energy source.

Energy source	Typical unit size	Variable[a]	Predictable	Dispatchable
Coal	600 MW	—	—	Yes
Nuclear	600 MW	No	Yes	No
Gas CCGT	Up to 500 MW	No	Yes	Yes
Gas open cycle	100 MW	No	Yes	Yes
Hydro with reservoir	Up to 500 MW	No	Yes	Yes
Pumped-storage hydro	Up to 500 MW	Yes	Yes	Yes
CHP	Up to 100 MW	Usually	Usually	No: because it is heat led
Energy crops and municipal solid waste	Up to 40 MW (at present in UK), in future larger units	No	Yes	Yes
Wind	Up to 8 MW	Yes	Not accurately	No
Landfill gas	1 MW	No	Yes	Yes
Run-of-river hydro	kW (in UK)	Yes	Not accurately	No
Photovoltaic	1 kW, domestic up to MW commercial	Yes	Not accurately	No
Wave	No commercial examples yet	Yes	Not accurately over long-term	No
Tidal barrage	No recent	Yes	Yes	No
Tidal stream	1.5 MW	Yes	Yes	No

a) Whether output varies in time.

as the primary energy resource is continuously available. However, their capability to be dispatched, which requires that their output can be changed automatically or at will by control engineers to follow the demand variations, depends crucially on their thermal/mechanical nature and differs from plant to plant as described in Chapter 3.

One characteristic shared by all the renewable sources excluding biomass and tidal, is their variability and relative unpredictability. This presents a challenge in integrating such sources in electrical power networks that have been designed to operate with traditional generators whose availability appears certain. In reality though, no plant can be completely available, there is always some probability of breakdown, the need for maintenance and so on. This topic is complex and often misunderstood. It is dealt with in some detail in Chapter 3.

2.10 Combining Sources

Many power systems at present make use of renewable sources, most commonly hydro and wind power. For now, it is sufficient to note, that, as mentioned earlier, geographical diversity of the wind resources assists integration. Geographical diversity also applies to the other renewable resources and is important to their integration into power systems.

The relationship between the different renewable sources, spatially and temporally, is also directly relevant to the issue of integration. Tidal current velocities are not correlated with wind speeds, and wave energy, although an integrated form of wind energy, is not strongly correlated with 10 minute or even hourly wind speeds. Solar power has its own unique pattern of variation, and biomass is flexible. How renewable sources of generation can be combined to meet load demand is a critical issue and is discussed in later parts of the book.

References

1 Star, F. (2007). Flexibility of fossil fuel plant in a renewable energy scenario: possible implications for the UK. In: *Renewable Electricity and the Grid* (ed. G. Boyle). Earthscan.

2 Killingtveit, A. (2013). Hydropower, Chapter 20. In: *Transition to Renewable Energy Systems* (eds. D. Stolten and V. Scherer), 381–401. Wiley-VCH.

3 Bardsley, W.E. and Choudhry, S. (2000). An approach to estimating hydro power system income gain from computerized water scheduling. *Nat. Resour. Res.* 9 (3): 215–222.

4 Liao, X., Zhou, J., Ouyang, S. et al. (2013). An adaptive chaotic artificial bee colony algorithm for short-term hydrothermal generation scheduling. *Int. J. Electr. Power Energy Syst.* 53: 34–42.

5 Roy, P.K. (2013). Teaching learning based optimization for short-term hydrothermal scheduling problem considering valve point effect and prohibited discharge constraint. *Int. J. Electr. Power Energy Syst.* 53: 10–19.

6 EWEA. Wind Energy Scenarios for 2013. A report by the European Wind Energy Association, August 2015.

7 Conradsen, K. and Nielsen, L.B. (1984). Review of Weibull statistics for estimation of wind speed distributions. *J. Clim. Appl. Meteorol.* 23: 1173–1183.

8 Seguro, J.V. and Lambert, T.W. (2000). Modern estimation of the parameters of the Weibull wind speed distribution for wind energy analysis. *J. Wind Eng. Ind. Aerodyn* 85: 75–84.

9 Loughborough University. CREST MSc Wind I notes. https://www.lboro.ac.uk/research/crest/research/windandwaterpower/ (accessed May 2019), 2007.

10 Sharpe, D. (2011). Aerodynamics of horizontal axis wind turbines, Chapter 3. In: *Wind Energy Handbook*, 2e (eds. T. Burton, N. Jenkins, D. Sharpe, et al.), 39–136. John Wiley.

11 Hansen, M.O.L. (2008). *Aerodynamics of Wind Turbines*. London: James and James.

12 Manwell, J.F., McGowan, J.G., and Rogers, A.L. (2010). *Wind Energy Explained: Theory Design and Application*, 2e. Wiley.

13 Sinden, G. (2005). *Wind Power and the UK Wind Resource*. Environmental Change Institute, University of Oxford.

14 WindEurope. The European offshore wind industry – key trends and statistics 2016; WindEurope, 2017.

15 The Carbon Trust. Offshore Wind: Big Challenge, Big Opportunity. The Carbon Trust, 2008.

16 Feuchtwang, J. and Infield, D. (2013). Offshore wind turbine maintenance access: a closed-form probabilistic method for calculating delays caused by sea-state. *Wind Energy* 16 (7): 1049–1066.

17 Duffie, J.A. and Beckmann, W.A. (2013). *Solar Engineering of Thermal Processes*, 4e. Wiley.

18 Green, M.A. (1982). *Solar Cells: Operating Principles, Technology, and System Applications*. Prentice Hall.

19 Nelson, J. (2003). *The Physics of Solar Cells*. Imperial College Press.

20 EERE. Energy.gov. Solar overview. Available at: http://www.eere.energy.gov/consumerinfo/pdfs/solar_overview.pdf (accessed May 2019) (2019).

21 Pitz-Paal, R. (2014). Solar energy – concentrating solar power, Chapter 19. In: *Future Energy*, 2e (ed. T.M. Letcher), 405–431. Elsevier.

22 Pinson, P., Reikard, G., and Bidlot, J.R. (2012). Probabilistic forecasting of the wave energy flux. *Appl. Energy* 93: 364–370.

23 Farrell, A., Pevin, R.J., Turner, B.J. et al. (2006). Ethanol can contribute to energy and environmental goals. *Science* 311: 506–508.

24 Department for Trade and Industry. Estimating Energy Requirements and CO_2 Emissions from Production of Energy Grasses, UK DTI report URN 01/797, 2001.

25 Sinden, G.E., 2007, DPhil thesis Environmental Change Unit, Oxford University

3

Power Balance/Frequency Control

3.1 Introduction

The essential function of an electrical power system is to meet the energy demand of consumers. This chapter describes the power demand characteristics of consumers and the technical and operational principles involved in the reliable delivery of power to them from a variety of conventional and renewable energy generators. The chapter is focused on technical issues, whereas market issues are addressed in Chapter 7; in particular, the detailed regulatory issues of control and operational requirements adopted in different countries where electricity systems have been liberalised.

3.1.1 The Power Balance Issue

Consider first the case in which the power system consists of a single generator driven by a prime mover and supplying a load. In Chapter 4, it will be shown that the frequency of the generated voltage is directly proportional to the rotational speed of the generator.

The generator has rotational inertia reflecting its spinning mass, and it is assumed that the prime mover is fitted with a *governor*. The function of a governor is to sense any changes in speed and to adjust the fuel supplied to the prime mover so that the speed (and therefore the frequency) is controlled.

The law of energy conservation requires that at any instant the power demanded by the load, referred to usually as *demand* or just *load* is supplied by the generator and/or by energy stored within the system. If the load is suddenly increased, the extra energy demand is initially supplied by the rotational inertia of the generator through a decrease of its speed. This decrease in speed is also reflected by a proportionate decrease in frequency. The governor senses this decrease in speed and increases the fuel supply to arrest the fall in speed and frequency. Exactly how far the frequency falls, how quickly it recovers and the frequency of new equilibrium state depends on the governor characteristics and the frequency dependence of the load (reflecting the fact that the demand of some loads, e.g. involving motors, will change with system frequency). All these issues will be discussed later in more detail. For now, it is sufficient to recognise the interdependence between system aggregate demand and power system frequency.

Complex power systems consisting of multiple interconnected generators supplying a large number of consumers respond much as the single generator but with all the generators and loads contributing to the system response during demand changes. Just as for a single generator supplying a load, the frequency of a complex interconnected system

Renewable Energy in Power Systems, Second Edition. David Infield and Leon Freris.
© 2020 John Wiley & Sons Ltd. Published 2020 by John Wiley & Sons Ltd.
Companion website: www.wiley.com/go/infield-renewable-energy

is the same at all parts of the network. For now, the idea of a unique system frequency should be accepted without question, but it will be properly explained in Chapter 4. A power system is, of course, never in equilibrium because the demand varies continuously as consumers switch on or off their loads. It can be concluded that these transient frequency shifts are an indicator of the imbalance between supply and demand at a particular instant. The power system frequency will drift downwards whenever demand exceeds supply, and vice-versa.

Conventionally, power systems are run so that their frequency remains within narrow bounds because:

- This ensures that electric motors directly connected to the mains (i.e. without power electronic converters allowing variable-speed operation) operate at virtually constant speed. A fixed speed is required in many consumer applications where an AC electric motor is used to drive a device at approximately constant rate; for example, a pump in a washing machine or a lathe in an industrial workshop.
- In electronic applications the mains frequency can be used as a basis for timing various processes.
- Transformers are sensitive to frequency variations and may be overloaded if the frequency drifts substantially from the nominal value.
- Finally, and most importantly, in traditional power stations the performance of the generators is dependent on the performance of all the auxiliary electric motor drives that deliver fuel and air to the boiler, oil to bearings and cooling services to several systems. If these auxiliaries under-perform due to low speed caused by low frequency, power station output can be reduced. As will be discussed later this phenomenon could lead to a run-away situation with cascaded shut down of power stations and black outs.

For a near constant frequency to be maintained, it is necessary that the supply of power accurately tracks the variations in demand. The causes and nature of electricity demand changes are the subject of the next section.

3.2 Electricity Demand

3.2.1 Demand Curves

Figure 3.1 shows the highly variable nature of the electricity demand over a day of a typical individual house in the UK, with a minimum demand of a few watts, an average between 0.5 and 1 kW and the maximum in the range of 5–10 kW, that is, 10–20 times the average load. Using a dedicated electricity generator to supply this house alone would be hopelessly expensive. The generator system would have to be large enough to meet the maximum demand, but most of the time it would be running at a very small fraction of its rated capacity. Any fuel driven prime mover operated in this way would be woefully inefficient. Energy storage, such as batteries, could improve generation efficiency but would have to be quite large, and therefore expensive. If each household in the UK met its own maximum demand of say 5 kW, 100 GW of plant would be needed for this sector alone.

Figure 3.1 Demand curve of an individual house.

3.2.2 Load Aggregation

The smoothing benefit arising from aggregation is of vital importance to electricity utilities. The more uncorrelated the demand among consumers, the more effective the overall smoothing. For a large power system this statistical effect is dramatic and is illustrated by the characteristic demand profile shown in Figure 3.2. This shows a typical demand curve over a day for the whole of England and Wales and should be contrasted with Figure 3.1. Not only is the peak to mean ratio considerably reduced but this curve is noticeably much smoother than that of the individual house. As a consequence, it is

Figure 3.2 Illustration of the smoothing effects of aggregation. Source: National Grid plc.

much easier to predict, and the generation required to supply this aggregate load can be scheduled and controlled very efficiently, as will be discussed later. The technical term used by the industry for the demand taking into account of this smoothing is demand after diversity. It is used widely in the design of electricity distribution networks where the demand is aggregated and system sized to meet the *after diversity maximum demand* (ADMD).

The value of interconnection to form large power systems should now be clear: it allows demand aggregation and the benefits that stem from this, primarily through the easier matching of supply and demand. Some proponents of renewable energy suggest that national grids will become redundant once generators are located near to consumers, but this is a misconception unless an unprecedented breakthrough in energy storage technology is achieved. Indeed, given the intrinsic variability of many dispersed renewable energy sources, interconnection may well prove to be even more valuable in the future.

3.2.3 Demand-Side Management – Deferrable Loads

Two important linked concepts are those of *deferrable loads* and *demand-side management*. A deferrable load consumes a certain amount of energy to provide a service but is flexible in terms of exactly when that energy is supplied because it possesses either an internal storage capacity or a large thermal inertia or because the consumer is flexible about the time when he or she requires the energy service.

Demand-side management (DSM) is a technique used by utilities to remotely regulate the demand required by deferrable loads so that their connection to the grid is scheduled according to the availability or cost of power. The use of off-peak electricity tariffs for night-storage heating is a familiar example in the context of the UK and helps to level out the demand difference between night and day so that a greater proportion of the generation can run at a constant output. Similar incentives are offered to commercial and industrial users, but, on the whole, consumers connected to the large reliable power systems found in affluent counties rarely give much thought to *when* they use electricity. For the most part, the generation in these systems is designed and controlled to meet the load, rather than the other way around.

Conversely, users of stand-alone electricity systems often time their use of electricity to suit the available generation. In fossil fuelled systems, it is usually a matter of levelling out the load. In systems relying mainly on wind or solar power, the timing of use is obviously related to the availability of resource.

Exploiting the deferability of loads is a useful tactic in any power system; it is especially valuable in systems relying on variable renewable energy sources, and can be far cheaper than employing energy storage. Many future energy scenarios that include a significant proportion of renewable energy assume widespread application of demand-side management to match supply and demand. Experience with DSM is presently limited and there is a pressing need for research to explore the extent to which consumers will be prepared to change their patterns of energy use. Research undertaken recently [1] has shown that for three quite different UK energy scenarios, DSM has a crucial role to play.

3.3 Power Governing

3.3.1 Power Conversion Chain

To understand the impact of renewable energy (RE) sources on power systems, it is essential to understand the way in which thermally driven generators are controlled. This is also of direct application to biomass fuelled systems, where in many cases the plant is similar to a standard fossil-fuelled plant.

Figure 3.3 shows the power flow diagram for a simple fuel fired generator [2], supplying electricity to consumers as depicted in Figure 1.12 in Chapter 1. Energy is first converted from chemical into thermal form in the boiler, from thermal into mechanical form in the turbine, from mechanical into electrical form in the generator and finally back to thermal, mechanical, light or chemical forms by the action of the consumers.

For the frequency to remain fixed, the fuel energy input to the boiler must be controlled to balance the variable consumer demand.

Importantly, the chain of power conversion units in Figure 3.3 contains two stages with intrinsic energy storage. The boiler consists of kilometres of piping that carry superheated high-pressure steam. Because of its large physical size, it constitutes a substantial thermal store; in fact, it contains enough energy to supply the turbine at full output for a few minutes. The turbo generator itself, because of its substantial rotating mass and its high rotational speed, also contains inertial stored energy, in this case only of the order of a few full-output seconds. These stored energies can be used to follow variations in demand over short periods. Rotational energy in the turbo generator is instantly available in response to decreases in system frequency. A sudden increment in demand will result in a frequency drop that will extract a part of the stored energy from the spinning mass. The more significant stored energy in the boiler is also available, with some delay, to supply any increments of electrical generation through the action of the governor.

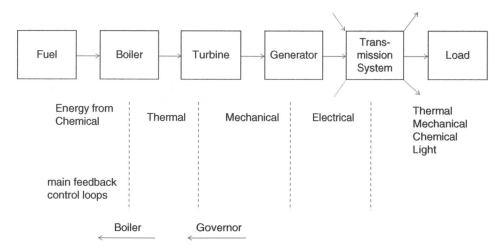

Figure 3.3 Power flow diagram for a thermal power station [2]. Source: Reproduced with permission of McGraw-Hill.

3.3.2 Governor Steady State Characteristics

The steam flow from the boiler to the turbine of a large conventional fossil fuelled generator is regulated by a valve of considerable size and weight. For the turbo generator to respond quickly to a requirement to increase or decrease its output, this valve has to be opened or closed as fast as possible. This is achieved through a hydraulic actuator. The control of the hydraulic actuator is provided by the governor. In the past governors were of mechanical nature but modern generators are fitted with electronic governors. The governor's function is to measure the rotational speed of the generator, compare it to the reference value (50 or 60 Hz) and based on the error signal, to instruct the hydraulic actuator to open or close the steam valve.

Figure 3.4 shows the action of the governor of a single generator supplying a load. Governors are designed to operate as proportional control systems, that is, an error must be present between the set point frequency and the actual frequency for the governor to alter the fuel or steam supply. This proportional control is characterised by the sloping lines in Figure 3.4, which has a fall or *droop* of 4% across the operational range. This 4% value has been found through extensive experience to be appropriate for stable governing and is widespread. Governors have the inbuilt facility that allows an operator to adjust the frequency at which the characteristic intercepts the frequency axis, known as the set *point*. For the line aa the set point is 52 Hz. With this set point, the system frequency is 52 Hz when the generator is supplying no load and is reduced incrementally to 50 Hz when the load increases to rated generator power. Changing the set point to 51 Hz moves the line to bb. Set point adjustment is of considerable importance because it allows power system operators to decide how the demand is shared by the generators on the grid, but more on this later.

It should be noted that although most plant employ governor droop control, it is possible with electronic feedback control to deliver *isochronous* governing where the frequency is held constant (i.e. there is no droop). This sort of control, which involves integral action, is commonly applied in small isolated power systems, perhaps supplied by a single diesel generator. With a group of generators, only one machine can be operated isochronously otherwise instability will occur if, say, two generators fight to control

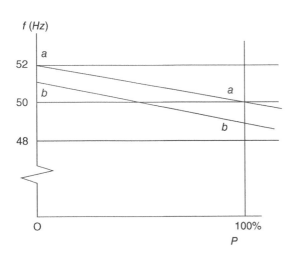

Figure 3.4 Frequency/power governor characteristic.

the frequency. Other smaller generators on the grid must be operated with droop speed control. Worked Example 3.2 illustrates the operation of such a small grid.

3.3.3 Parallel Operation of Two Generators

When two generators are connected in parallel and are jointly supplying the demand in a small power system, the load is shared according to the set points of their governors. The best way to illustrate this sharing mechanism and the resulting system frequency is by means of an example.

In the small power system of Figure 3.5a two generators A and B rated at 50 and 100 MW, respectively, supply a load of 100 MW. Both generators are fitted with governors having a droop of 4% and a no-load set point of 52 Hz. Lines aa and bb show the frequency-power (f–P) characteristics of the two generators.

The division of load between the generators when they are supplying 100 MW will now be determined. Let line CC located Δ_f below 52 Hz represent the frequency at which this sharing takes place.

Using trigonometry $\frac{P_A}{\Delta_f} = \frac{50}{2}$ and $\frac{P_B}{\Delta_f} = \frac{100}{2}$ therefore $\frac{P_A}{P_B} = \frac{1}{2}$, however $P_A + P_B = 100$ MW, therefore $P_A = 33.3$ MW and $P_B = 66.6$ MW.

In this case, the demand is shared in proportion to the generator ratings.

To determine the system frequency: from $\frac{P_A}{\Delta_f} = \frac{50}{2}$ $\Delta f = 33.3 * 2/50 = 1.33$, hence the frequency at which line CC intercepts the f-axis is $f = (52 - 1.33) = 50.67$ Hz.

Suppose that generator A is more modern and therefore more efficient in terms of fuel used per generated kWh. We wish to use the set point adjustment of the governors to arrive at a new more economic load-sharing condition that maximises the contribution from generator A.

One way of implementing this is to shift the set point of generator B so that its characteristic is displaced from bb to BB with the result that each generator is supplying 50 MW. The system frequency is now 50 Hz and the new set point of generator B can be found from $\Delta f/50 = 2/100$. Hence, $\Delta f = 1$ and the new set point is 51 Hz.

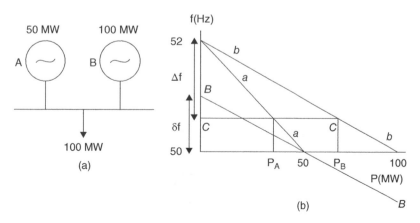

Figure 3.5 Parallel generators. Source: Reproduced with permission of McGraw-Hill.

3.3.4 A Multi-Generator System

The two-generator load-sharing paradigm can be extended to explain the behaviour of a typical power system where a very large number of generators are interconnected through a transmission network to feed a multiplicity of loads. If all generators have the same standard governor droop characteristic and a common set point, then they will all share load in proportion to their ratings. In fact, for the economic reasons illustrated in the example of Figure 3.5 generators with the lowest cost of energy production are run with their governor set points adjusted at full output (e.g. characteristic bb in Figure 3.5). Such generators are incapable of providing increments in power; therefore, their governors are unresponsive to demand increases. A number of generators that are less economic to operate, or for other reasons to be explained later, are run with governor set points that results in them operating at part load. Such generators are capable of responding to demand increases. It can be shown that in such well-integrated systems and to a good approximation all the generators with *active governors* can be lumped together into one very large generator having a rating equal to the sum of the ratings of the constituent generators and possessing the typical 4% droop due to the combined governor action of all the machines. This *equivalent generator* has a rating several times the value of the typical rating of generators on the network. The reason for this will be explained later. The f-P characteristic of such an equivalent generator is shown by aa in Figure 3.6 where f is the system frequency at which the present steady demand P_d on the equivalent generator is supplied. Because of the large size of this equivalent generator, aa intercepts the P far to the right. Suppose that an additional generator is connected to the network the 4% droop characteristic of which is given by bb. At the system frequency f, the additional generator injects into the system a power ΔP. Power balance now demands that the system frequency rises by an amount such that the power injection by the equivalent generator is reduced by ΔP. The figure shows that this requires a very small frequency rise Δf. It can be concluded that in a large interconnected power system, an increase or decrease of injected power from a single generator will have only a small effect on frequency.

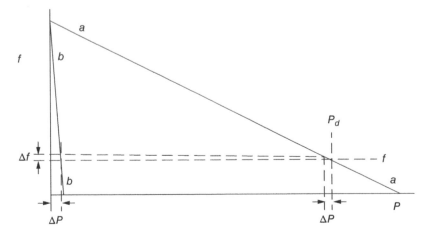

Figure 3.6 Frequency-power characteristic of an infinite bus.

A power system that is fed by active governor generation capacity overwhelmingly larger than the rating of a single generator is known as an *infinite bus*. The frequency of such a system is taken to be independent of the power injected by a single generator. In Figure 3.6 this 'independent' frequency is given by f. Therefore, the power injected by each individual generator is defined by the intersection of its characteristic with the frequency line f. By shifting line bb parallel to itself, that is, adjusting the governor set point, the power injected by each generator can be regulated to the desired level.

3.3.5 The Steady State Power–Frequency Relationship

The new frequency at which a power system will stabilise after a sudden change in generation or demand depends on:

a) the combined effect of the droops of all active governors;
b) the total load on the system and its frequency dependence.

The small-increment frequency-power behaviour of a power system can be described approximately through a simple mathematical relationship [2].

a) When the frequency changes by $-\Delta f$ (a frequency drop due for example to a sudden demand increase), governor action increases the power output of generators by, say, ΔP_G.
b) Consumer demand consists of a variety of loads. All resistive loads, that is heating and lighting as well as electronic devices, such as PCs, are, in general, insensitive to small frequency variations. However, the large majority of electrical drives, whether in household appliances or most importantly in industry, are frequency sensitive unless they are connected to the mains through a power electronic interface. A large induction motor drive will run marginally slower at lower frequency, provide less mechanical power and therefore absorb lower electrical power. Let us define the decrease in total system demand by $-\Delta P_D$ for a drop in frequency $-\Delta f$.

It can now be seen that for a drop in frequency $-\Delta f$, the effective increase of available power ΔP is partly due to the increase of power from generators and partly due to the *released demand* from loads. Therefore,

$$\Delta P = \Delta P_G + \Delta P_D \text{ for a frequency change} - \Delta f.$$

If it is assumed that the small variation ΔP is linearly related to the frequency change Δf, the power–frequency relationship is therefore given by

$$\Delta P = -K\Delta f \qquad (3.1)$$

where K is a constant expressed in MW/Hz and known as the *power–frequency* or *frequency response characteristic*, or *system gain*. The negative sign in the equation indicates that a drop in frequency results in an increase in the net power available within a system. This relationship gives a rough value of the resulting system frequency after the initial transients have subsided in the case of a sudden major loss of generation or less usually a sudden increase in demand. The relationship describes the situation maybe a minute after the incident (see Figure 3.9 discussed later); therefore, K is not related to rotating inertia effects. The value of K for a specific power system depends on the governing characteristics of the generators and the combined frequency-power sensitivity of the loads. Its value varies during the day as the share of load may shift between frequency sensitive and insensitive consumers.

The value of K can be estimated by tripping out a large generator and measuring the resulting frequency drop after the transients have subsided. Preferably, values of K can be derived at random times from the records of system response after the loss of a generator due to a fault. To ensure stable operation after the most severe contingency that the utility considers as *credible*; that is, probable, K should be such that the system frequency does not drop below a value that leads to serious underperformance of power station auxiliary drives. Of course, the utilities cannot protect the system from contingencies that have very low probability; for example, the simultaneous loss of two large power stations. If such events were to occur, measures are in place to maintain the integrity of the system by means of unavoidable power interruption to some consumers, but more of this later.

3.4   Dynamic Frequency Control of Large Systems

3.4.1   Demand Matching

Figure 3.2 shows the daily variation of the aggregate load for a typical utility. The pattern of low load at night and the peaks in the morning and afternoon is a feature of such demand curves. Superimposed on this daily cycle are faster random variations.

Power systems operation is conventionally broken down into different time scales ranging from seconds to days. During second to minute load variations partly loaded plant respond through governor action. Generating plant responding in this time scale are said to be providing *continuous service* or *frequency response*.

The next time scale is *load following*, which involves the connection/disconnection of plant to balance the anticipated load increases/decreases. This time scale covers approximately 10 minutes to several hours, during which decisions are taken in response to the trend in demand on the basis of plant operation economics. For example, in anticipation of the early morning increase in demand, the system operator either directly or through bidding mechanisms (Chapter 7) is responsible for ensuring that sufficient capacity is available to meet these large but relatively slow and predictable changes in demand.

Fossil fuelled generators and especially plant fuelled by coal, can be readily made to follow demand. Such generators are brought on-line to supply any escalating demand according to cost effectiveness and flexibility. This results in the layering of generation shown in Figure 3.7. Nuclear power being inflexible occupies the bottom of the pile. Combined cycle gas turbine (CCGT) generation and efficient coal occupies the next layer. These two layers supply what is known as the *base load* with other coal fired generation and imports from adjacent networks required to do the load following. Pumped hydro, when available, is used to extract energy when demand is low and deliver it during peak demand.

For load following to be effective it is essential to have accurate predictions of the demand curve. Programming generation at this time scale is known as *unit commitment* or *generation scheduling*.

Figure 3.7 Demand curve of the England and Wales power system. Source: National Grid plc.

The reader should not misunderstand the nature of the layering in Figure 3.7. For example, it would be wrong to assume that a constant load like office lighting is supplied from a nuclear power station while a toaster is supplied from coal! Rather a power system could be likened to a bathtub in which water is fed from all the generators and is extracted by loads. The level of water in the bathtub would represent the frequency of the system that the operators endeavour to keep constant. Some generators, for example nuclear, provide a constant inflow of water. Other generators are instructed to supply water when the level is detected to be falling. In such an analogy there is no way one can tell which 'generated water molecule' reaches which load. In other words, electrons cannot be labelled.

3.4.2 Demand Forecasting

Accurate forecasts of demand are required because:

- Electrical energy cannot yet be stored economically.
- The largest proportion of generating plant is thermal in nature. An unfortunate feature of this plant is the considerable delays involved in preparing the cold generators for connection to the power system (several hours) and the restrictions in the rate at which a steam driven turbo generator can be loaded after connection. These operational delays are dictated by the thermal/mechanical safety requirements of massive boilers and of turbo generator sets.
- Thermal generators using steam turbines have an upper limit of power generation equal to their *nameplate rating* but also a lower limit dictated by cavitation problems in the turbine blades at low throughputs of steam. Consequently, when a turbo generator is connected to the network it should be loaded to a level at least equal to the minimum recommended by the manufacturers (from 30 to 50% of rated power).

Figure 3.7 shows that there are periods during the day (e.g. 6–7 a.m.) when the rate of demand growth is considerable. To maintain system frequency, the injected power must closely track the trajectory of the demand curve. Unfortunately, because of the sluggishness of the thermal plant, this tracking cannot be done unless preparative action is taken some hours before the event.

It may be concluded that there is an absolute necessity to carry out a *demand forecasting* activity in order to prepare and progressively load plant as required. Utilities have invested considerable effort in forecasting the daily pattern of demand. Through years of experience they have evolved sophisticated mathematical techniques to correlate demand to the aggregate of the national habits and to other factors such as weather. All methods are essentially based on the fact that demand exhibits regular patterns. Forecasting techniques adjust past demand to present weather and other conditions. Meteorological data on temperature, wind speed, humidity, cloud cover and visibility are used as variables because such factors have an important bearing on heating and lighting demand. The art of load forecasting has been refined to such an extent that estimates are rarely in error by more than ±3% and on average in the UK system they are accurate to within ±1.3%.

Demand prediction techniques are constantly being refined but there will always be occasions when unforeseen circumstances increase, or depress the load. The average daily errors in demand in a typical month on the English system are shown in Figure 3.8; during this period, the maximum error in prediction was just under 4% and on average it was less than 1.53%. The standard error during this month was 1.6% and, as the average demand was about 32 GW, this corresponds to about 300 MW. The figure shows that on 11 November the forecast was adrift by 3.5% representing a maximum error of over 1 GW.

In conclusion, even if all generation were 100% reliable, a substantial reserve would still be required because it is simply not possible to predict the demand on a power system exactly.

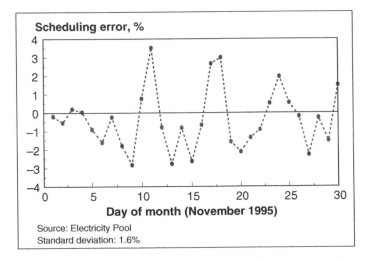

Figure 3.8 Typical scheduling errors on the network in England and Wales. Source: Reproduced from Reference [38] with permission of Earthscan.

Figure 3.9 Operational and statutory limits of system frequency for the UK power network. Source: National Grid plc.

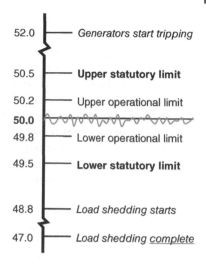

3.4.3 Frequency Limits

Although perfect balance between supply and demand is not achievable, the balance has to be good enough so that under normal operating conditions, the frequency does not drift outside prescribed operational limits.

Figure 3.9 shows that for the UK, statutory limits on permissible frequency variation are ±1% (95% of week) with absolute limits in emergencies of +4% and −6%.

However, the National Grid's usual operational limits are ±0.4%, which are tighter than the legal requirements. Under unusual circumstances, for example substantial loss of demand, the frequency may rise uncontrollably and generating sets are fitted with over-frequency protection that will trip them off the system if the frequency reaches 52 Hz. At the other extreme, due to loss of generation or less frequently, unexpected demand increase, the frequency may fall below its statutory limit.

Large groups of non-critical loads (i.e. excluding hospitals, emergency services etc.) are fitted with staggered under-frequency protection so that switching off such loads; that is, *load shedding* begins at about 48.8 Hz and is completed by 47 Hz. Further drops in frequency could trigger cascade tripping of power stations. Such cascade tripping incidents are not unknown, several having taken place in developed countries during the last decade. Complete shutdown of the electricity supply or blackout is an occasional but traumatic event for millions of affected citizens trapped in lifts, underground trains and so on. Full restoration of supply may take as long as several days.

The topic of frequency control is of particular importance in systems with substantial renewable energy penetration.

3.4.4 Generation Scheduling and Reserve

Demand forecasting provides a fairly accurate picture of the expected load over the following 24 hours so that enough generators are scheduled to provide the expected demand plus a *reserve*, a concept to be discussed in the following section. The process is complicated by the disparate characteristics of plant on the system.

In Chapter 7, it will be shown how generators are loaded on economic grounds subject to various constraints. In anticipation of an increase in demand, a choice has to be made by a utility on which uncommitted generating sets from a number of available generating units (some sets are not available as they are undergoing repairs or maintenance) have to be prepared (heated, synchronised and loaded) and eventually shut down when the demand declines. This is a complicated techno-economic choice. In a privatised energy market, one company, usually the one in charge of the transmission system and known as the *system operator* (SO), has the role of plant scheduler, but here decisions are taken in the context of the contractual relations between the different participants in the electricity market. Chapter 7 describes how regulatory tools, tariffs and bidding systems are used in these circumstances to ensure that supply tracks demand as required.

3.4.5 Frequency Control at Different Timescales

The maintenance of frequency involves *response* from generators over several timescales ranging from seconds to days [3, 4]. This response is not only required to follow the demand's fast variability from second to second and its slow variability over the day, but also sudden substantial mismatches between generation and demand; for example, during system faults.

Figure 3.10 illustrates a typical system frequency trajectory plotted on a non-linear time scale. From the origin of the graph until about 8 s, the frequency exhibits the usual noise associated with minor mismatch between the continuously varying demand and the efforts of generation to match it through governor action. At 8 s, something unpredicted and serious takes place. Let us say that a large power station trips because the overhead line connecting it to the transmission system suffers a mechanical failure due to high winds and accumulated snow. This *contingency* results in an instantaneous large shortfall of generation. The trace describes a typical time history of the frequency and

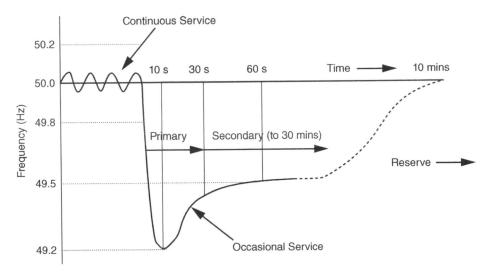

Figure 3.10 Frequency response requirements. Source: National Grid plc.

the measures taken to constrain the frequency excursion within the statutory and operational limits. Such measures are taken by all utilities but are given a variety of labels. Here, the labels shown in Figure 3.9 are the ones adopted in the UK.

A power system has at its disposal a number of generators with diverse characteristics. These are arranged in a hierarchy of plant appropriate for operation at different timescales as described next.

Continuous or *frequency response* is provided by generators equipped with appropriate governing systems that control their outputs to counteract the frequency fluctuations that arise from relatively modest changes in demand or generation [3]. Large generators on the grid are selected on technical and economic merits and instructed by the system operator to operate in frequency-sensitive mode (i.e. under active governor control) to provide this service. For this to be achieved, some generators are held below maximum output.

Occasional service or *reserve* is available to contain significant and abnormal frequency excursions caused by sudden mismatches in the generation/demand balance (e.g. loss of generation) [3]. Part-loaded large synchronised generators as well as deferrable loads fitted with frequency-sensitive relays provide such services. The reserves are subdivided into *primary* and *secondary* categories. The initial rate at which the frequency drops after the incident at 8 s in Figure 3.10, is controlled and limited instantly by the inertial energy release from all the decelerating generators (and consumer motors) on the system. This provides a breathing time before the fast governors of some generators begin to act. Massive steam valves have to be opened hydraulically and increased steam flow has to be transported from the boiler to the turbines. Therefore, it takes a finite time for the substantial stored energy in the boiler to be exploited.

Primary reserves require the most rapid generator response. The key requirement for generators allocated this task is that they should be capable of increasing their active power output within 10 s of predefined system frequency excursions and be capable of maintaining this response for a further 20 s [3].

Secondary Reserves requires a slower initial response but maintained for longer periods of time. This requires the capability of increasing the active power output within 30 s and maintaining the response for a further 30 min [3]. Fast response capacity is also provided by partly loaded hydro or pumped storage (when available) that are not bedevilled by the constraints imposed on thermal plant. Water driven plant can respond in a few minutes and be started up automatically when the frequency falls below a critical value.

A fundamental feature of generators providing frequency response and reserves, collectively known as *operating margin*, is the requirement for generators to have *headroom* in order to increase output. Operating margin is the difference between available generation and actual demand. Generators providing such services will therefore be part-loaded. Besides hydro and pumped storage plant (if available), the main providers of such services are flexible large coal fired power plant.

Beyond primary and secondary reserves, power systems have further robust *tertiary* defences known as *standing reserves*. These are sourced from unsynchronised standby generators capable of mains connection and generation of the instructed level of output within 20 min [4]. Typically, standing reserve is provided from generators driven by open cycle gas turbines and reciprocating internal combustion engines.

High frequency response services are required in the event of excessive system frequency events when large loads are suddenly lost. Such services happen rarely and are initiated through governor action requiring to either reduce output or cease generation altogether.

Finally, utilities have some control over the demand by implementing *voltage control*. System voltage, like system frequency is rarely exactly at its nominal value but is allowed to vary within controlled limits. One response to a loss of generation, which may occur due to manual intervention or automatically, is a reduction of system voltage. Total system load will fall with voltage depending on the nature of the loads. Voltage and frequency reductions can cope with serious credible demand or generation changes, although only in exceptional circumstances would both be allowed to reach their minimum or maximum values.

3.4.6 Meeting Demand and Ensuring Reliability

In the early power systems, a single generator provided power to a local network that serviced a number of loads. In such a system, the generator would have to be exceedingly flexible and capable of supplying the lowest to the highest demand. In the case of failure of this generator, the supply would be interrupted unless a similar generator was kept in reserve providing a 100% *back-up*. By interconnecting such isolated systems and allowing exchange of power, the risk of generator failure can be shared so that lower back-up capacity is required to provide the same level of system reliability.

As mentioned earlier, system balancing requires a number of reserves to deal with unexpected demand fluctuations and loss of generation. These reserves are determined on a statistical basis that takes into account the margin of error in demand prediction and the probability of system failures. The aim is to meet specific levels of reliability that historically have proved to be acceptable to customers. To maintain near 100% reliability would be prohibitively expensive. Critical loads such as hospital operation theatres and major computing networks are further backed up by *uninterruptible power supply* (UPS) systems based on batteries or local internal combustion engine generators.

To maintain frequency stability, the *operating margin* or *balancing reserve* provided must be sized in relation to three factors:

1. The maximum deficit that may suddenly occur in a power system due to loss of generation
2. The expected availability of all plant connected to the system and
3. The demand prediction error.

In the UK system, the Transmission System Operator (TSO) has traditionally carried a balancing reserve of 2.5 GW [5], although in recent years this plant margin has been falling. Within limits, the larger the available headroom of the frequency response plant the greater the reliability of the power system; however, the presence of these additional part-loaded generating sets on the system significantly increases the system fuel costs. Any part-loaded machine has a poor efficiency due to *standing losses*: fixed losses incurred by a generator when running unloaded.

The risk of demand being unmet is measured by a statistical quantity known as *loss-of-load probability* (LOLP). In the UK pre-privatised system, a LOLP of 9% was considered acceptable; that is, expected loss of supply incidents for each customer in nine winters per century. The LOLP gives no information on how much load will be shed or for how long. It is understood that reliabilities of the same order are expected in present privatised systems of developed countries although figures are difficult to obtain. For such a reliability to be maintained in addition to the short-term reserves discussed earlier, a system or *plant margin* must be maintained. This is the difference between *installed capacity*, including imports and exports, and maximum annual *peak demand*. This margin is necessary because at any time there are bound to be plants that have broken down or are out for maintenance. For reliable operation, a power system must have a plant margin that is substantially greater than the operating margin. Taking as example a system with 70 GW winter peak demand, to guarantee that the demand will be met to a typical reliability expected in an industrialised nation about 84 GW of conventional capacity (20% above peak demand) should be available on the network. It should be noted that at present the UK margin is much less than this, reflecting a lack of planning and investment rather than the operational acceptability of such limited spare capacity.

In privatised or *deregulated* power systems all these activities are managed through the complex financial instruments that define the contributions and obligations of several independent generation providers all feeding into the same network. For example, in the UK and in other similarly privatised networks, system operators pay considerable premiums to power generators who provide response and reserves. However, the issue of ensuring sufficient installed capacity has recently been attracting attention since the present market arrangements in the UK and the USA seem not to be delivering this. Some argue that a 'capacity market' should be created, although there are other approaches being explored such as enhanced rewards for delivering power system services likes reserves in comparison with bulk generation (see e.g. the US utility ERCOT). Chapter 7 will deal in more detail with this important issue.

3.4.7 Capacity Factor and Capacity Credit

When renewables displace significant amounts of conventional generation plant, extra conventional plant capacity margin is required to maintain system supply reliability. It is important to recognise that additional plant margin will only be required at times of high electricity demand and low input from variable sources. At other times, additional margin will not be required as there will be sufficient conventional plant available to meet demand irrespective of the contribution from variable sources.

The capacity factor of a generator [5], as mentioned in Section 2.4.3, is usually defined as the ratio of its yearly energy output to the output it would have produced if it operated continuously at its nameplate rating. Due to unavailability caused by maintenance schedules, breakdowns and so on, no plant achieves a capacity factor of unity. Base-load thermal generators have a capacity factor of 85–90% when new, declining over the years until they are decommissioned. Wind turbines achieve capacity factors of 20–50% depending on the windiness of the site. In the UK, the average capacity factor of onshore wind farms is ~30% with offshore schemes achieving higher values. Often, uninformed commentators quote the figure of 30% when they claim that wind turbines require back-up for 70% of the time. In fact, wind turbines provide some power between cut-in and cut-out wind speeds for 80% of the time. The capacity factor does not determine back-up requirements that must be assessed statistically.

Statistical analysis confirms that, for limited penetration, the capacity factor is a good guide of the probability that the generator in question will be available to contribute towards meeting demand [6]. On this basis variable sources are incapable of providing the same level of reliable or *firm* power during demand peaks as conventional generators but they are still capable of providing a contribution to this. This firmness is known as the *capacity credit* and is a measure of the amount of load that can be provided by variable plant with no increase in the LOLP. This can be illustrated by an example. A 1 GW dispersed installation of wind farms with a 30% load factor will provide the same yearly energy output as a 350 MW gas fuelled plant of 85% load factor. Both plant statistically have a capacity credit of 300 MW, assuming as in this case, that the wind penetration into the power system is at a low level.

These calculations assume that the availability of generated power is completely uncorrelated to the demand. Although this assumption is correct for gas-fuelled plant, this is certainly not so in the case of the wind plant. As will be shown in the next section, due to aggregation of wind farms and the characteristics of the wind resource, there is a positive correlation between the availability of wind power and the demand in the UK. In the main this is because wind speeds in the UK are higher in the winter when the loads are also high. This would not be true in parts of the world where electricity load is dominated by air conditioning. In these locations it would tend to be solar power that correlates positively with load.

3.5 Impact of Renewable Generation on Frequency Control and Reliability

3.5.1 Introduction

The introduction of variable RE generation into a network will have an impact and incur associated costs in two main categories [5]. The first can be labelled as *balancing impact* and relates to the management of demand fluctuations from seconds to hours. The second, referred to as *reliability impact* relates to the requirement that there is enough generation to meet the peak demand. Both balancing and reliability involves statistical calculations. The introduction of variable RE generation introduces additional uncertainties that can be quantified in terms of operational penalties that have to be taken into account when the value of electricity from RE sources is calculated.

There is a widespread, but mistaken belief that operation of an electricity system with renewables causes serious problems. A common misconception is that significant additional plant must be held in readiness, to come on-line when the output from wind plant ceases. This would indeed be true in an island situation, with, as an example, wind being the principal source of supply. Modest amounts of variable renewables within an integrated electricity system pose, however, no threat whatsoever to system operation. The reason for this is that these amounts do not add significantly to the uncertainties in predicting the generation to ensure a balance between supply and demand. So, the risk of changes in the output from variable renewable sources has only a small influence on the needs for reserves.

In the following sections, we will limit our discussion mainly to the integration issues of wind power because this is currently the non-schedulable renewable energy source making the largest impact, and is likely to remain so for the foreseeable future. In a later section, the impact from other renewables will also be briefly considered. The discussion will review the temporal availability and operational aspects of wind power, and the penalties incurred for different penetration levels so that the system reliability is maintained at the desired level.

3.5.2 Aggregation of Sources

In Section 3.2.2, it was shown how integrated electricity systems benefit immeasurably from the aggregation of consumer demand. Fluctuating sources can benefit in the same way. Figure 2.9 in Chapter 2 confirms the benefits of adding the output of geographically dispersed wind sites in the UK [7]. Aggregation here has provided its usual benefits by smoothing the output over short and medium timescales.

3.5.2.1 The Monthly Distribution of Power Availability
The seasonal wind power availability from dispersed sites in the UK shown in Figure 3.11 indicates limited production during summer and greater than average production during winter [7]. On average, twice as much electricity is generated during the winter compared to the summer months. This pattern matches the seasonal demand pattern in the UK.

3.5.2.2 The Daily Distribution of Power Availability
Figure 3.12 shows that, on average, wind power availability is higher during the daytime than at night in the UK [7]. This trend is present irrespective of the time of year and is of benefit in a system where the demand peaks during the afternoon period when the wind power availability is near its maximum.

Figure 3.11 Variation in monthly wind power output. Source: Reproduced from Reference [39] with permission of Environmental Change Institute, Oxford University Centre.

Figure 3.12 Average daily variation in wind power output. Source: Reproduced from Reference [39] with permission of Environmental Change Institute, Oxford University Centre.

3.5.2.3 Short Term Variability

The variability of wind power will cause changes in the power generated from one hour to the next. The maximum expected rate of change from hour to hour provides an indication of the reserves required to deal with shortfalls in supplying demand. Wind speed variations within the 15–25 m/s band will result in no change of power as the wind turbine will be operating at full output for winds in this range. However, variations within the band 4–15 m/s will result in substantial power changes. The degree of dispersion of the resource will again be of advantage as increments of wind at one site will be compensated by decrements at another.

Figure 3.13 shows the benefit of geographical dispersion from an EWEA study [8]. In the graph, the frequency of occurrence of positive and negative changes observed in the hourly values of wind power output from actual wind farm measurements is shown for a single wind farm and for wind power aggregated over the whole of Germany. Whereas a single wind farm can exhibit hour to hour power swings of up to 60% of installed capacity this figure is less than 20% for aggregated wind farms. These maximum changes are likely to occur about once a year. These figures are significant because they indicate the requirement of fast response part-loaded thermal plant to *ramp up* or *ramp down* and this would have economic implications as discussed later.

3.5.2.4 The Capacity Factor

Figure 3.14 shows the results of a UK study on the relative capacity credit between seasons and between onshore and offshore resource as a function of wind power penetration [8, 9]. Due to the windiness of the UK offshore areas, the capacity credit is higher than 60% during winter decreasing to less than 25% in the summer for small penetrations. The onshore picture is similar but scaled down. As expected, the capacity credit decreases with increasing penetration. It may be concluded that, in the UK, because of

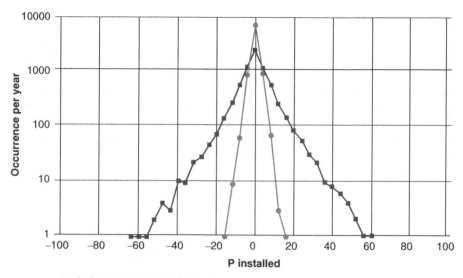

wind power aggregated over Germany

single wind farm, inland Germany

Figure 3.13 Frequency of occurrence of hourly wind power variations. Source: Courtesy of EWEA [8].

- Offshore Capacity Credit (%) winter
- Offshore Capacity Credit (%) summer
- Onshore Capacity Credit (%) winter
- Onshore Capacity Credit (%) summer

Figure 3.14 Seasonal capacity credit as a function of penetration for onshore and offshore UK wind resource. Source: Courtesy of EWEA [8].

the favourable weather conditions, a statistical correlation exists between availability of wind power and demand. The capacity credit associated with wind power and therefore the contribution of wind towards reliability could be substantially higher than that calculated from the annual average wind capacity factor.

3.5.3 Value of Energy from the Wind

The financial benefit of wind power can be calculated by determining the cost of supplying the total demand and then subtracting the cost of supplying the residual demand (after wind power is added) from the existing or a reduced mix of generation. In practice, this calculation is extremely difficult to do and, for simplicity, many analysts concentrate on fuel savings that are more straightforward to estimate. *Ideal fuel savings* are simply those calculated from the cost of displaced generation and do not take account of any changes to operation forced on conventional plant by the time-varying characteristic of the wind plant. To estimate the fuel savings more realistically the following operating issues should be considered, [10]:

- If wind generation is subtracted from the gross demand (wind power can be viewed as negative demand) the residual demand will have more variability than the gross demand. This means that the output of fossil fuel plant providing continuous or frequency response service will have to be adjusted more frequently and to a greater extent. Additionally, with wind generation there is a need to ensure that the system can respond adequately to unpredicted changes over longer time periods. Extra balancing reserves are provided by greater headroom but the lower loading level of thermal plant that this requires would result in lower operational efficiency of thermal plant. Finally, at higher penetrations (above 20%), some thermal plant may have to be shut down and started up to maintain adequate reserves. This will incur what are known as cycling costs. All these effects make up the *balancing impact*.
- Wind generated electricity may increase the size of the system margin required to maintain the required level of reliability. The reason for this is that wind plants are less likely than thermal plants to be available to contribute towards times of peak demand. This effect makes up the *reliability impact*.
- At high penetrations, energy may need to be *spill*, *discarded* or *curtailed* because for operational reasons this energy cannot be safely absorbed while maintaining adequate reserves.

The issues that define these impacts and the associated costs and penalties that have to be assigned to wind generated electricity are discussed in what follows.

3.5.4 Impact on Balancing

Thermal plant incur some costs when they are run in frequency control mode, It can be shown [6, 10–12] that if variations in a variable source and in demand occur roughly independently, the total resulting variation in the net load to be met by the thermal plant is approximately a 'sum-of-squares' addition of the components:

$$\begin{pmatrix} \text{Total variability} \\ \text{of load on thermal} \\ \text{units} \end{pmatrix}^2 = \begin{pmatrix} \text{Total variability} \\ \text{of electricity} \\ \text{demand} \end{pmatrix}^2 + \begin{pmatrix} \text{Total variability} \\ \text{of variable} \\ \text{source} \end{pmatrix}^2 \qquad (3.2)$$

Thus, for example, when the average power variation of the added source equals that of the demand itself, the total variability is not doubled but increased by 40%. This has some important implications. The impact of fluctuations in variable sources at low penetrations is practically zero; they are lost as noise among demand fluctuations.

Over longer time scales, the level of operating reserve required at any given time depends on two key factors; uncertainties in demand prediction, and the probability of loss of the largest generation plant on the network. When wind power plant is introduced into the system, an additional source of variation is added to the already variable nature of demand. To analyse the additional variation caused by the wind plant it is important to appreciate that the requirement is that the entire system must be balanced instead of balancing each individual load or resource. The operator has to ensure that the average system reliability is maintained at the same level it would have been without the wind resource.

However, the crucial question is by how much does wind generation increase the balancing uncertainties? Intuitively, it is known that minute-to-minute fluctuations in wind output are largely uncorrelated to load. This implies that the additional uncertainty introduced by wind power does not add *linearly* to the uncertainty of predicting the load. As for the issue of variability dealt with before, it can be shown that when errors in predicting the output from variable sources occurs independently of those in predicting demand, the combined error is a sum-of-squares addition [10–13].

$$\begin{pmatrix} \text{Average error in} \\ \text{predicting net load} \\ \text{on thermal units} \end{pmatrix}^2 = \begin{pmatrix} \text{Average error in} \\ \text{predicting electricity} \\ \text{demand} \end{pmatrix}^2 + \begin{pmatrix} \text{Average error in} \\ \text{predicting variable} \\ \text{input} \end{pmatrix}^2 \quad (3.3)$$

Demand prediction techniques are constantly being refined but there will always be occasions when unforeseen circumstances push up, or depress the load. Equation (3.3) indicates that for small penetrations of variable sources the prediction errors are lost among load fluctuations. However, since demand is fairly predictable, forecasting errors from substantial penetration of wind will incur some penalty.

Analysis of the combined uncertainties of wind, demand and conventional generation based on the sum-of-squares calculation of Eq. (3.2), make use of the standard error in predicting the generation/demand balance. On typical developed country networks, 1 h ahead, this averages at around 1% of the demand. For 4 h ahead, this figure rises to 3%.

The need to schedule reserve to cover for possible trips of conventional thermal plant emphasises the point that no generation is 100% reliable. While the loss of a typical 1000 MW of thermal plant is a real risk, it is unlikely that 1000 MW of wind plant would be suddenly lost unless this is the result of loss of cable connection to very large future offshore wind farms.[1] It is also assumed that dispersed wind plant is not sensitive to a common mode disturbance; that is, the plant rides through voltage dips caused by faults on the transmission system. The more wind that is installed, the more widely it is spread and sudden changes of wind output across a whole country simply will not occur in practice, [14].

Calculations can be made over various timescales to determine the need for extra reserve. Figure 3.15 shows the estimated additional balancing power needed (expressed as a percentage of installed wind power) as a function of wind power penetration [13]. At 20% penetration, 7% of extra operating reserves are required if *persistence*[2] in the wind is assumed. For perfect forecasting, only 2% additional capacity would be needed.

---

1 In practice, multiple cables are likely to be used to limit the largest likely single loss of generation.
2 'Persistence' wind forecasting assumes that the wind power output 1 h ahead is the same as at time zero. 'Perfect' forecasting as in Figure 3.15 assumes that wind power can be predicted with total precision.

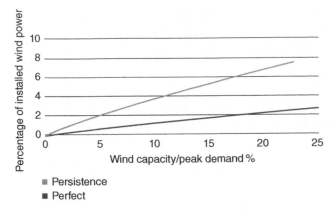

■ Persistence
■ Perfect

Figure 3.15 Additional balancing power for wind. Source: Reproduced from Windpower Monthly, December 2003 [13].

Although this work was completed some years ago, the results remain broadly correct and persistence forecasting remains in use as the benchmark against which to assess improved forecasting methods.

Clearly these back-up figures are modest, but what about the associated costs? In the UK, the SO has estimated that 10% wind penetration would increase balancing costs by £40 m a year, which is equivalent to £0.002/kWh. A 20% penetration will increase the cost to £0.003/kWh. These figures should be viewed in terms of the retail cost of electricity, which at the time of writing is in the region of £0.14/kWh to £0.18/kWh. Hence, a 20% penetration in the UK would incur a 3% additional cost on electricity at present prices.

For relatively little expenditure the predictability of wind could be greatly improved. This would be accomplished partly through the installation of extra weather data monitoring stations (e.g. anemometry towers a few tens of kilometres from major wind farms) and partly through sophisticated computational techniques. Programmes to provide enhanced predictability are being developed in several countries [15]. A 2011 review summarises the state of the art of wind forecasting [16]. This is an active research field, and steady progress can be anticipated.

Figure 3.16 illustrates this by showing a typical 1-h wind forecast using sophisticated techniques against actual output for one wind farm over a period of a week. Such techniques can provide considerable cost benefits in operating reserve and are now in widespread use by system operators. Wind power forecasting has improved since this work was completed; indeed, it remains an active area of research, nevertheless the results would look substantially the same.

3.5.5 Impact on Reliability

In addition to operating reserve, some system margin in excess of system peak demand is required and this will be affected by the level of wind penetration. The example of a power system with a peak demand of 70 GW will be given based on [7]. To guarantee long-term security to an accepted level of reliability, utilities in general would require that about 20% (14 GW) of that peak must be additionally available on the network

Figure 3.16 Wind farm forecast (+1 hour) versus actual output, Ireland, 2004. Source: Reproduced with permission of Garrad Hassan and Partners Ltd.

as system margin. The system in question has a yearly energy demand of 350 TWh. If 10% of the system energy were to be generated by wind power, it is possible to assess what should the new plant margin should be. To generate 10% of the energy, that is, 35 TWh, a wind power capacity of 35 000/(8736 × 0.3) = 13 MW is required, where 8736 is the number of hours in the year and 0.3 is the average wind farm capacity factor. The 35 TWh would be generated by 35 000/(8736 × 0.75) = 5.3 MW of conventional generation with a yearly load factor of 0.75. Ideally the wind plant would directly substitute for this conventional plant, however the variability of wind power means that not all of this conventional plant can be removed from the network. Statistical calculations by utilities indicate that 81 GW of conventional plant would be required to meet the same level of reliability. So only 3 GW (84–81 GW) can be removed from the network, that is, 2.3 GW (5.3–3 GW) is retained because of wind power variability; that is, 17% of the wind power installed capacity.

In the UK context of a planned 20% penetration from wind and to maintain system reliability, the associated cost lies within 0.03 and 0.05 per kWh.

3.5.6 Discarded/Curtailed Energy

As the capacity of the variable sources injected into a system increases, there might be occasions when the available power from such sources cannot be used. If the penetration is substantial, there might be periods when the available power from renewables exceeds demand, or cannot be accommodated by the transmission or distribution system. However, even before this stage is reached, energy from variable sources will have to be shed because the power system would need to keep a minimum level of thermal plant generating to maintain adequate operating reserve.

Discarding energy from variable sources (sometimes called *dispatch-down*) poses no particular operational difficulties. Output from wind turbines can be controlled through blade pitch variation, from photovoltaics through inverter control and from hydro, wave and tidal schemes through similar control techniques. However, this discarding or *curtailment*[3] of energy results in an economic penalty on variable sources, which becomes increasingly important at high penetrations. This penalty is difficult to assess as it depends heavily on the flexibility of the base-load units, that is, the extent to which they could be operated in a stable regime at low power and upon how rapidly their output could be increased if required. For the level of penetration expected over the next decade or so, the penalties due to discarded energy are unlikely to be of major significance. It should also be noted that curtailment may also be required due to distribution or transmission system constraints. Reference [5] states that five out of six studies show that at a penetration of 20%, curtailed energy ranges from 0 to 7%. Most studies are showing that with sensible design, curtailment due to local network capacity limitations would be rarely required for a wind penetration of up to 20%. This is supported by recent data from Ireland [17] showing an overall curtailment of 2.1% of the wind energy potential in 2012. During the year in question, 14.4% of the electricity demand of Ireland and Northern Ireland was provided by wind. For the Irish power system five different reasons are given for curtailment:

i) System stability requirements (synchronous inertia, dynamic and transient stability)
ii) Operating reserve requirements, including negative reserve
iii) Voltage control requirements
iv) Morning load rise requirements
v) System Non-Synchronous Penetration limit (currently 50%).

There is concern that as wind generating capacity increases further, so will the levels of curtailment unless other steps are taken to increase the operational flexibility of the power system. The All Ireland Grid study report specifically included the expected curtailment in 2020 [18]. Without major new investment in energy storage, or the wide scale deployment of demand-side management, it was concluded that curtailment could rise to over 14% with 9 GW of wind.

3.5.7 Overall Penalties Due to Increasing Penetration

In previous sections, operational penalties due to increases in variable source penetration were reviewed. A number of studies have been carried out to provide estimates of these penalties as the penetration of renewables increases. The majority of these studies relate to wind power, as this is the variable source with the largest installed world capacity to date. Table 3.1 give some indicative figures based on EU-funded studies and on Danish and UK thresholds linked to operational experience of wind farms. The additional costs are tabulated in terms of the level of penetration.

3.5.8 Combining Different Renewable Sources

The benefits of combining different variable sources were mentioned in Section 2.10. As the capacity of a non-dispatchable source increases, its marginal value declines, primarily because successive increments of capacity are correlated with those already on the system. In contrast, combining capacity from different renewables with uncorrelated or complementary outputs can be of considerable benefit [10, 11, 19].

---

3 The term curtailment is sometimes used for dispatch-down undertaken for system-wide reasons, whereas the term constraint is used when local (distribution of transmission) limitations are the primary cause.

Table 3.1 Implications of increasing wind energy supply on the UK network.

Wind power penetration	Measures required	Cost penalties
Up to 5%	None	Negligible
5–10%	Occasional instances when some energy from the wind is discarded and more part loading of thermal plant required	0.1–0.2 p/kWh
10–20%	As before, plus more use of pumped storage or hydro to balance wind power	0.2–0.8 p/kWh
20–50%	May be necessary to build more storage, or peaking plant, or retain old coal plant, depending on relative costs. (Extra storage will benefit system as a whole)	>0.8 p/kWh

Typically, a combination of wind and solar could be beneficial. In some circumstances, thermally driven winds can be strongest after sunset, so that the combination of wind and solar usefully covers periods of high demand. Other studies indicate that a combination of wind and tidal (two sources having statistical independence) increases their value compared with the case of having more of the same [10].

A UK study [19] indicates that a contribution from a mix of PV solar and wind plus domestic CHP has the potential to reduce significantly the overall variability that would have been experienced if only one renewable technology were to provide the total contribution.

The potential synergies among different renewable sources are clearly much too important to ignore, and they may often make the combined exploitable potential larger than the sum of parts considered in isolation.

3.5.9 Differences Between Electricity Systems

It is appropriate here to stress that results from studies on one particular network do not necessarily apply elsewhere [5]. The operational viability and costs of integrating renewable energy depend on a number of factors that characterise the local resource as well as the structure of the electricity network. These factors include:

- The strength and temporal variability of the resource;
- The possibility of geographical dispersion over a large area to gain the advantages of aggregation;
- The possible complementarity between different types of renewable resources;
- The correlation if any between availability of the resource and demand variation;
- The extent to which the magnitude of the resources can be forecast; some weather patterns are more predictable than others;
- The robustness of the electricity network and the proximity of transmission lines to the areas of maximum resource availability;
- The transmission links, if any, to adjacent networks;
- The operating practices of the network, in particular how far in advance the system balancing reserve is planned;
- The type of conventional plant in the network. Smaller and more modern thermal plants are more flexible than large base-load plant such as nuclear.

3.5.10 Limits of Penetration from Non-Dispatchable Sources

Early on in the development of renewables, the UK's Central Electricity Generating Board (CEGB) carried out a number of extensive simulation studies to estimate the impact of large penetrations [10, 11]. To illustrate the challenges, an extreme example from one such study is reproduced here. Figure 3.17 shows the output that would have been expected from 25 GW of dispersed wind capacity (middle graph) and 10 GW of tidal (lower graph) alongside the demand over a period of one month. In this example, 30 and 13% of the consumer energy would be supplied from wind and tidal, respectively. The wind power varies less rapidly than demand, tidal more rapidly. Figure 3.18 shows the effect of subtracting the output from wind and tidal power from the demand, leaving a residual load (dashed curve) to be met by conventional thermal plant. Although this is an early study (and the CEGB was ended decades ago with privatisation of the sector), the figure does illustrate rather well the challenge of high renewable energy penetrations.

More recently, studies [20, 21] have considered the possibility of meeting the entire demand from a mix of renewables. Figure 3.19 taken from Streater's work, [20], shows hour-by-hour variations in the time variable renewables. Penetration levels of such magnitude result in periods when the available power from the RE sources exceeds demand. Even before this stage is reached, for reasons of system reliability, the RE sources would have to be curtailed. Table 3.2 shows the capacities if the different renewables installed and their penetration, defined as the output divided by the total load (i.e. ignoring any curtailment). It is also apparent, that for the year in question a deficit occurs between weeks 15–18 and weeks 44–52. In principle, the shortfall could be made up from biomass-based generation.

Denmark, and less so Germany, generate a high percentage of their total electricity needs from wind power and are planning further capacity. A recent study from Elkraft, a Danish system operator, asserts that a wind penetration up to 50% is technically and economically feasible. This is based on an increase of installed capacity of 3.1 GW in 2005 to 5 GW in 2025 and a substantial expansion of the Danish grid. The study claims

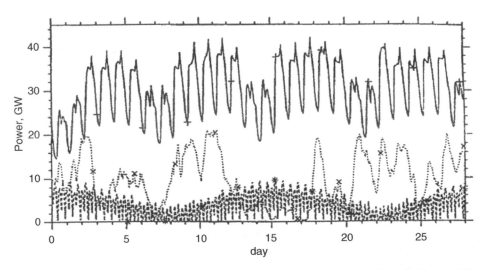

Figure 3.17 CEGB simulation study of large penetration. Source: Reproduced from Reference [10] with permission of IET.

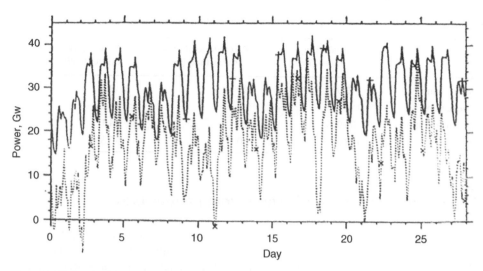

Figure 3.18 Residual demand to be met by conventional generation. Source: Reproduced from Reference [10] with permission of IET.

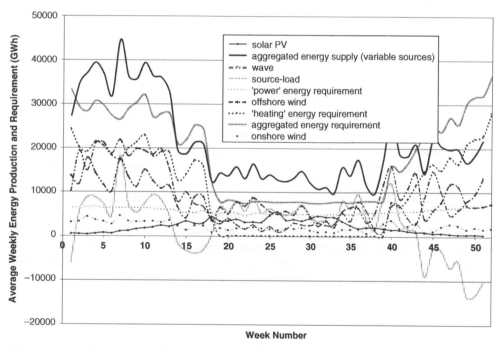

Figure 3.19 Hour by hour variation in renewable energy generation over one year, compared with variations in energy requirements.

that even with this penetration, it will be necessary to shut down wind turbines only for a few hours in a year. Similar studies by the DENA in Germany [22] indicate that up to 20% penetration by wind is possible with minor extensions of the grid and no need for construction of additional power stations.

Table 3.2 Assumed renewable energy capacities and production in the Streater study [20].

	Annual energy produced (TWh)	Percentage of total variable load (%)	Mean power output (GW)
Onshore wind	72 000	20	8.2
Offshore wind	144 000	40	16.4
Solar photovoltaic	36 000	10	4.1
Wave	108 000	30	12.3

3.6   Frequency Response Services from Renewables

With the anticipated rise in the penetration of variable renewables, power systems will be required to accommodate increasing second-to-second imbalances between generation and demand requiring enhanced frequency control balancing services. In principle, renewable generation may contribute to frequency regulation services but, depending on the nature of the frequency support and the plant, this could require headroom in the form of part loading. Technologies that could potentially provide frequency support services are biomass, water power, photovoltaics and variable-speed wind turbines. Chapter 7 will show that economics dictate that energy from renewable sources should generally be used as fully as possible whenever available. Although this seems to contradict the idea of part loading such plant, there are some occasions when priorities may dictate otherwise.

With large penetrations from renewables there will be occasions, for instance during low demand days over summer, when the number of conventional generators needed to supply the residual load will be so few, that adequate level of response and reserve may be difficult to maintain. Moreover, the displacement of conventional generators by renewable energy sources can result in a reduction in the total power system inertia, and this makes control of system frequency more difficult. Under such conditions, renewable generators could be unloaded and instructed to take part in frequency regulation, and also to mimic inertial response. Such a provision has been made, for example, in the Irish (ESB) code, for the connection of wind turbines. In a privatised system the opportunity benefits of running in these modes must more than compensate the loss of revenue from generating at less than the maximum potential. How different renewable generation sources could contribute to power system frequency control is explored in more detail next.

3.6.1.1   Wind Power
Early wind power technology was mainly based on simple fixed one- or two-speed stall regulated wind turbines with little control over the dynamic performance of the generator. These turbines did, however, add to the total inertia of the power system. However, in recent years, pitch regulated variable-speed wind turbines have been developed capable of increased conversion efficiencies but also of substantial control capabilities. In principle, modern wind turbines are capable of providing continuous response by fast increase in power from part loading through blade pitch or rotor speed control in response to

drops in frequency and through the same mechanism provide high frequency response through fast reduction in power in response to increases in frequency.

As wind power capacity has increased, to the extent that at times it is the dominant form of generation in parts of Denmark and Northern Germany, there is an increasing demand for wind capacity to be dispatchable and to behave more like conventional generation. Very large wind farms are now expected to conform to connection standards that limit ramp rates for increase in power and also to contribute to frequency regulation under times of high network stress. These requirements are increasingly included in the national *grid codes* that regulate access to the public networks. In these early days it is unclear to what extent this will result in wind power being curtailed, for example to comply with given ramp rates, and to what extent such constraints add value to the system operator.

Conventional steam generation plant assist the network frequency stability at the onset of a sudden imbalance of demand over supply by slowing down. Wind turbines respond differently. The stored energy is in the rotor inertia and fixed speed turbines will provide a limited benefit from their inertia provided that the voltage and frequency remain within their operating limits. Variable-speed wind turbines will not normally provide this benefit as their speed is controlled to maximise the energy production from the prevailing wind.

Large wind turbines are now almost always of the variable-speed type and as they increasingly displace conventional generation the total system inertia from such generation will decrease. Consequently, the rate-of-change of frequency and the depth of the frequency dip caused by a sudden loss of generation will both increase. Sophisticated control arrangements to support system functions are likely to be requested by utilities as wind penetration increases. The wind turbine control can be augmented to rapidly increase power in response to a rapidly falling frequency that would result from a sudden loss of a major conventional generator. In effect, energy stored in the wind turbine rotor is extracted and supplied to the power system, to be made up when the power system has recovered from the frequency transient. This response is sometimes referred to as *synthetic inertia*. A number of wind turbine manufacturers now offer turbine controllers that can deliver such inertial response but there remain uncertainties as to how this would benefit power system operation in practice. Early research on inertial response from wind turbines concentrated on the modification of the turbine controller and for simplicity assumed constant wind speed. Figure 3.20 shows that the modification to the turbine controller is in principle straightforward.

Real inertias respond to the rate of change of frequency, df/dt, and so to create an inertial response it is sufficient to add to the normal turbine torque demand (T_{ref}) an additional signal proportional to df/dt (H being the magnitude of the desired synthetic

Figure 3.20 Modification of wind turbine torque controller to provide inertial response.

inertia constant). More recently, research has been undertaken to explore the expected impact on the power system as a whole. Work on the Irish power system [23, 24] shows that the number of wind turbines generating any time determines the expected inertial response, and this is related to the aggregate power output from wind at the time in question, although there remains a significant uncertainty as to how many turbines may be generating. Meanwhile, a study of the UK [25] has made use of a probabilistic method developed initially to take account of varying wind speeds across a wind farm [26], and combined this with simplified system frequency response model (similar to that described in Section 3.7) and a representation of wind speeds across different regions of the UK. It is concluded that wind capacity can make a useful contribution to reducing the initial rate of fall of system frequency and increasing the minimum frequency before recovery occurs. Figure 3.21 shows an example frequency response following a loss of 1.8 GW of generation on the UK system. It should be noted though that synthetic inertia cannot exactly replace actual physical inertia as there are delays in the frequency sensing and the implementation of the power control. These should be minimised as far as possible without compromising accuracy, but are nevertheless significant in the context of the timescales of frequency transients.

As described earlier in this chapter, continuous regulation of frequency under normal system operation is provided by governor droop. Such a service can in principle be provided by wind turbines but this requires de-rating to provide operational headroom. With modern variable-speed pitch regulated wind turbines this can be provided in two distinct ways: by adjusting the blade pitch away from the optimal operating point; or by operating at a tip-speed ratio to one side of the maximum C_p (see Figure 2.8). These two options are presented in Figures 3.22 and 3.23, respectively, both taken from [27]. In Figure 3.22 it can be seen that below rated wind speed the rotor speed increases linearly with wind speed in the normal maximum power tracking mode, but a small non-optimal

Figure 3.21 Comparison of frequency transient with and without inertial response from wind.

Figure 3.22 Use of pitch control to provide 10% operational headroom.

Figure 3.23 10% operational headroom achieved through over-speeding of a wind turbine.

blade pitch is used to reduce power output by 10%. Above rated, the pitch increases with wind speed to limit power to 90% of the normal rating. The 10% headroom created in this manner can be accessed quickly by returning the pitch control action to normal. An alternative to using pitch control to de-rate the turbine, is to operate consistently away from the optimal tip-speed ratio. The control illustrated by the dashed line in Figure 3.23 deliberately over-speeds the turbine by an amount calculated to reduce power output by 10%. The advantage of over-speeding rather than under-speeding is that there is increased energy stored in the wind turbine rotor and this allows such droop control to be combined with inertial response more effectively.

Finally, grid codes require wind turbines to maintain power infeeds to the system even under transient local voltage reductions. Such reductions are usually due to fault conditions in the vicinity of the wind farm. It can be shown that maintenance of power infeed from all generators is essential to ensure system recovery after a fault clearance.

3.6.1.2 Biofuels

Traditional thermal plant could be described as *capacity limited*; that is, capable of theoretically generating its rated output continuously as gas, coal, oil or fissionable material is abundantly available on demand. In contrast, energy crop-based [12] plant could be described as *energy limited* because the locally harvested fuel is limited in nature and may or may not be capable of sustaining all year-round continuous plant generation at full capacity. This depends on whether transporting biomass fuel from remote areas is economical.

Biomass plant would be expected to operate as a base-load generator running as far as possible at full output. Such plant would be able to contribute to continuous low or high frequency response services similarly to conventional plant. For a low frequency response, the plant would need to run part-loaded; a convenient strategy providing extra income if, say, due to a low crop yield year the stored fuel would not be capable to service continuous full output.

Land filled gas plant size is in the range of 0.5–1.5 MW and because of their small size they would not generally be suitable for the provision of frequency regulation services.

3.6.1.3 Waterpower

Small and medium sized hydro schemes without significant storage capacity are characterised by substantial variability of output depending on rainfall. Because of this and their small size they are not suitable for frequency regulation duties.

Tidal schemes could be very large indeed and their output would be highly predictable. Such plant would incur exceptionally high up-front capital costs and long pay back periods. Operation revenue is vital to service the large loans and it is unlikely that frequency response revenue based on part loading would be attractive enough. As such schemes are not yet in the planning stages, the jury is still out on their frequency control capabilities.

The comments made previously on wind power generally apply with reservations to future wave and tidal stream power schemes. As the technology is still in its infancy and commercial schemes are not yet in existence, it is not known how their dynamics may be capable of responding to signals derived from frequency deviations.

3.6.1.4 Photovoltaics

Here, a distinction should be made between large concentrated PV installations and numerous roof top systems. For large installations, the comments on wind power apply albeit with some reservations. As the PV systems are interfaced to the grids through power electronic converters and as no mechanical inertia is involved, the speed of response in increments or decrements in power flow can be very fast indeed. The

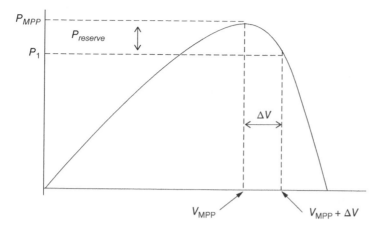

Figure 3.24 PV power voltage characteristic showing de-loading to create operational headroom.

larger the PV system the slower its dynamic response, although even large systems are exposed to passing clouds and can present rapid changes in system output.

As with other renewables, in the future, PV systems may be required to operate at part load thus providing 'headroom' for continuous or 'occasional' frequency control.

Unlike wind turbines, PV systems have no physical inertia, so any frequency response requires deliberate operation below the maximum power point. Figure 3.24 shows that in principle this is straightforward: the operating point is shifted by an amount ΔV from the voltage for maximum power, V_{MPP}, thereby creating the required operational headroom, $P_{reserve}$.

At this stage of the PV technology, roof installations in most countries are not yet numerous enough to provide a credible frequency response service. However, in years to come if, as predicted, costs plummet and installations are numbered in millions, it is conceivable that the local inverters are fitted with sophisticated controllers to assist system frequency stability. Since PV systems are very capital intensive, there is an economic driver to operate these at maximum output; the rewards for de-loading would have to be significant to change this approach.

3.7 Frequency Control Modelling

3.7.1 Background

Simple lumped parameter modelling is able to provide a rough estimate of the effect on frequency control of feeding relatively large amounts of renewable energy into a power system. A system model of this sort is outlined next [28], together with results showing the dynamic impact of high penetration of wind energy, and also the way in which dynamic control of demand can be used to good effect.

3.7.1.1 Modelling a Generator

The dynamics of real generator sets are highly complex and differ considerably between sets. However, it has been shown that a governor with a droop characteristic can be usefully modelled as a proportional controller [29].

The first step is to calculate the generator's target power output, P_{TAR}, using the 4% droop characteristic:

$$P_{TAR} = \left(\frac{f_{SP} - f}{0.04 \times f_{NOM}} \right) P_{MAX} \tag{3.4}$$

where f_{SP} is the generator's set point in Hz, f is the current grid frequency, f_{NOM} is the nominal grid frequency and P_{MAX} is the generator's capacity in MW. The next step is to proportionally reduce the error over time between P_{TAR} and the actual output, P at time t, using:

$$P_{(t+dT)} = P_{(t)} + (P_{TAR} - P_{(t)}) \cdot G \cdot dT \tag{3.5}$$

where G is the governor gain. It can be shown that an appropriate value for G is about 0.3 as this results in a realistic settling time in frequency of the order of 15–20 s after a step-change in load.

To get approximate but useful results it has been shown [29] that the total amount of primary response on the system can be modelled by a single governor-controlled generator of sufficient size to represent all generators with headroom. Also, the total amount of base generation can be modelled by an additional very large generator, but on fixed full output.

3.7.1.2 Modelling Released Demand
Many loads on the grid consist of rotating machines. As mentioned in Section 3.3.5, there is a built in frequency dependence caused by the fact that these machines slow down as the frequency drops, and thus consume less power. It has been found empirically that for the UK the total active power demand decreases by 1–2% for a 1% fall in frequency depending on the *load damping constant, D* [29]. This change in power is the released demand. It is treated in the simulation as an injection of active power, P_R, given by:

$$P_R = -D \cdot P_L \cdot \left(\frac{f - f_{NOM}}{f_{NOM}} \right) \tag{3.6}$$

where D for the UK is assumed here to be 1.0 and P_L is the total load that would exist if there were no built-in frequency dependence.

3.7.1.3 Modelling the Grid's Inertial Energy Store
As already stated, the grid frequency falls as all the spinning machines on the system begin to slow down. In effect, the demand deficit is being met by extracting energy from the rotational inertia of all the generators (and spinning loads). The fall in frequency will continue until the demand deficit is met by a combination of released demand and increased generation due to governor response.

In the simulation, all the inertia is assumed to be stored in an equivalent single flywheel of moment of inertia, I, rotating at grid frequency, $\omega = 2\pi f$ rad/s. The total kinetic energy stored is therefore:

$$KE = \frac{1}{2} \cdot I \cdot \omega^2 \tag{3.7}$$

The inertial storage capacity of a power system is measured by an *inertia constant, H*, which is the number of full-output seconds of energy stored (assuming nominal frequency).

Therefore,

$$HP_{GMAX} = KE = \frac{1}{2} \cdot I \cdot \omega^2$$

where P_{GMAX} is the total generation capacity. H typically is within the range of 2–8 s [29]. For this study, H is assumed to be 4. The inertia I for the system is calculated once at the start of the run:

$$I = \frac{2 \cdot P_{GMAX} \cdot H}{\omega^2_{NOM}} \tag{3.8}$$

For each step of the simulation, the total power surplus, P_S is then calculated:

$$P_S = P_G + P_R - P_L \tag{3.9}$$

where P_G is the total generation, P_R is the released demand and P_L is the load. Clearly, P_S is the power going *into* the inertial energy store. Given that for each simulation time slice, dT, energy must be conserved, then:

$$KE_{(t+dT)} = KE_{(t)} + P_S \cdot dT \tag{3.10}$$

Hence:

$$\frac{1}{2} \cdot I \cdot \omega^2_{(t+dT)} = \frac{1}{2} \cdot I \cdot \omega^2_{(t)} + P_S \cdot dT \tag{3.11}$$

which provides a difference equation for calculating the new frequency for each step of the simulation:

$$\omega_{(t+dT)} = \sqrt{\omega^2_{(t)} + \frac{2 \cdot P_S \cdot dT}{I}} \tag{3.12}$$

The simulation then is carried through the following steps for each time slice, dT:

1. Calculate P_L by summing the connected loads
2. Calculate P_G by summing the total generation
3. Calculate P_R using Eq. (3.6)
4. Calculate P_S using Eq. (3.9)
5. Calculate the new ω (and f) using Eq. (3.12)

3.7.2 A Modelling Example

The model just described was used to assess the effect that a large wind power input would have on the frequency stability of a power system. The example simulated is extreme and has been chosen because it illustrates key issues.

Wind speed data from 23 UK sites were used in the simulation. A 50-h period containing exceptional wind variability was chosen so as to provide a major challenge to integration. It was assumed that the variation in wind speed and physical separation of the wind turbines in each site would smooth second-to-second variations in power and thus the power system could be adequately modelled on a minute-to-minute basis.

For each site, the power output was calculated on the assumption that a wind farm comprising 150 4-MW variable-speed wind turbines was present at each site. A purely cubic power-wind speed relationship was assumed with a cut-in wind speed of 2 m/s, a rated wind speed of 15 m/s and a cut-out speed of 25 m/s. The output power from the

Figure 3.25 Simulated wind power using measured wind speed data from 23 UK sites.

23 sites were added together to give a total maximum generation capacity of 13.8 GW. This represents a level of penetration of 25% as a fraction of peak demand.

The simulation results are shown in Figure 3.25. The maximum power reached in the 50 h chosen was 4.6 GW and the minimum was 550 MW. During the period, the largest sustained drop in wind power occurred during the 37th hour when 5.5 GW of wind was lost in 4.5 h.

A simulation was set up with this wind generation connected, along with 3170 MW reserve (providing both primary and secondary response) and a frequency dependent load of 36 000 MW at 50 Hz. The level of reserve was chosen arbitrarily so that the maximum frequency excursion would remain approximately within the statutory limits. Enough base generation was provided such that the reserve was half-loaded when the wind power was at its average for the 50-h period – approximately $36\,000 - 2500 - (3170/2) \approx 32\,000$ MW.

Wind predictions were not used and the reserve requirements are large enough to compensate for substantial but untypical large positive or negative variations of the wind resource over several hours. The resulting grid frequency-time relationship for this simulation is shown by the curve 'without DDC' in Figure 3.26.

Between hours 20 and 25, the wind speed increased substantially to the extent that at 24 hours the base load plus the input from the wind exceeded the nominal demand. The consequence was that the frequency increased excessively to a level determined by the frequency dependence of the load. This undesirable situation could have been easily avoided by fitting wind turbines with a frequency dependent power regulator that reduces power generated at higher frequencies.

At the opposite extreme between hours 40 and 42, the output from the wind farms was so low that the base load plus the total available reserve (the remaining half of 3170 MW) was insufficient to balance the demand and the frequency had to drop to a level determined again by the frequency dependence of the load. This undesirable situation would have been prevented if at 37 h use was made of meteorological data predicting a power decline from the wind. Standby generating plant would have then been commissioned to provide the expected shortfall in generation.

Figure 3.26 Simulation of frequency with wind on the system.

3.8 Energy Storage

3.8.1 Introduction

Energy storage devices capable of being topped up during periods of low demand and discharged during periods of high demand would aid enormously the frequency control problem. Pumped storage schemes are classical but very expensive examples of such devices. Storage devices, if suitably cheap and efficient could be of benefit to the integration of high levels of RE source penetration, although if such devices were to be present in a power system, they would be of operational benefit to traditional energy sources as well.

Income may be derived from an energy store by charging it when the local electricity value is low and discharging it when the value is high, but the level of income critically depends on the round-trip efficiency and how this compares with the system electricity cost differential. If, at times, the grid at the point of connection of the embedded renewable generation cannot absorb the entire output of the generator a local storage device could prevent curtailment. Whether this is an economically viable strategy depends, among other things, on the capital cost, the round-trip efficiency and the operation and maintenance costs of the device. Another source of income from such devices results from the supply of ancillary services; for example, reactive power, voltage and frequency control and emergency power during a power outage. It is clear that assessing the economic benefit of storage is not a simple matter and that if renewable energy sources have low capital costs it may be more effective to curtail output whenever a surplus exists, rather than attempt to store the surplus.

3.8.2 Storage Devices

As mentioned, storage technologies depending on their characteristics may aid the integration of renewables, and they could assist the operation/control of a power system over the range of timescales discussed in Section 3.4.1. Conventional storage technologies include:

1. *Large hydro* is an 'old' renewable but whenever available could facilitate the take up of variable renewable sources by suitable timing of water release.
2. *Pumped hydro* is an excellent energy storage technique as far as the maturity of technology is concerned but few attractive sites exist and upfront capital costs are very high. Nevertheless, it is widely viewed at the most cost-effective means of large-scale electricity storage.

   Less conventional technologies include:
3. *Compressed air storage* has been talked about for some time and involves the storage of compressed air in disused underground cavities; for example, exhausted salt mines. At present, it is uncertain how economically and technologically viable this technology is.
4. *Hydrogen* can be produced by the electrolysis of water using energy from a renewable resource. It can then be 'burnt'[4] as fuel to generate electricity. Alternatively, it can be piped as a gas or liquid to consumers to be used locally providing both electricity and heating in a total energy scheme, or being used for transport. The combustion of hydrogen results in energy plus pure water with no harmful emissions or by-products. This may not be attractive if electricity is the final product since the round-trip efficiency is very low (generally below 50%). For this reason, there is much interest in using the hydrogen for transport, but this depends on much improved on-board storage systems.
5. *Flow Cells* operate in a mode similar to that of a car battery but without involving the electrodes. Instead, when the Flow Cell is used as a 'sink', electrical energy is converted into chemical energy by 'charging' two liquid electrolyte solutions. The stored energy can be released on discharge. In common with all DC systems connected to the AC network, a bi-directional power electronic converter is required.
6. *Batteries.* The lead-acid battery is one of the most developed battery technologies. It is a low cost and popular choice of electrical energy storage but has disadvantages in terms of energy density per unit weight, short cycle life, and the dependence of the delivered energy on the rate of discharge. As a consequence, a large variety of other batteries have been under intensive development to provide high round trip efficiency, low life-cycle cost, high reliability, capability of deep discharge and large number of charge-discharge cycles, low maintenance, high power energy density per unit weight and low capital cost. Although they are generally perceived as too expensive for general inclusion in large power systems, high temperature batteries such as the Zebra cell [30] are of increasing commercial interest at sub-station level. Lithium ion batteries have become the preferred batteries for use in electric vehicles and as such their performance has improved considerably and their costs have fallen. Consequently, although still expensive for power system applications, they are now being tried out. A good review of lithium ion batteries can be found in [31].

---

4 Literally burnt in an engine, or oxidised in a fuel cell.

7. *Flywheels.* There has been a number of projects carried out to store energy in very fast rotating flywheels. The major problems that had to be overcome were the maintenance issues and losses relating to bearings, the low specific strength of standard materials such as steel and the associated severe failure management problems at high speeds. Contemporary flywheels are made of fibre-reinforced composites that have powdered magnetic material introduced into the composite that when magnetised form either the rotor of a high-speed motor/generator or the rotating element of passive magnetic bearings. The flywheel motor/generator is interfaced to the mains through a power electronic converter. Currently the technology is expensive and only used for niche applications.

8. *Super Capacitors*, alternatively known as ultracapacitors, consist of a pair of metal foil electrodes, each of which has an activated carbon material deposited on one side. The activated carbon sides are separated by a paper membrane and then rolled into a package. Ultracapacitor operation relies on an electrostatic effect whereby charging and discharging takes place with the purely physical (not chemical) and reversible movement of ions. As a result, there are some fundamental property differences between ultracapacitors and battery technologies including long shelf and operating life as well as large charge-discharge cycles up to 500 000.

9. *Superconducting magnetic energy storage* (SMES) stores energy within a magnetic field created by the flow of direct current in a coil of superconducting material. Typically, the coil is maintained in its superconducting state through immersion in liquid helium at 4.2 K within a vacuum-insulated cryostat. A power electronic converter interfaces the SMES to the grid and controls the energy flow bi-directionally. With the recent development of materials that exhibit superconductivity closer to room temperatures this technology may become economically viable.

10. *Heat or cold store.* There is a long tradition of using thermal storage to assist in power system operation. The UK's old white metre off-peak tariff, and the more recent Economy 7 tariff were both primarily for charging storage heaters. More recently, sophisticated ways to use heat and cold storage in the context of power systems, has been explored. This technology involves the modulation of the energy absorbed by individual consumer electric heating elements and refrigeration systems for the benefit of overall system power balance. An example of this is described in the next section. Heat storage can also improve the characteristics of solar thermal power stations making their power output to a great extent independent of the instantaneous solar radiation.

Many excellent reviews of energy storage are available, and with particular emphasis on the benefits to electricity systems. Interested readers are directed to References [32–37]. Whether wide scale deployment of energy storage is used in future power systems to facilitate increased penetrations of renewable energy will depend on the alternatives. These are the demand-side management of the loads, which is discussed in more detail next, and increased interconnection of power systems since this allows the benefits of the increased temporal diversity of widely geographically dispersed generation. Evaluation of which approaches, or combinations of approaches, are most likely to deliver the greatest benefits at the lowest costs is a complex issue, and at present there is no consensus among power system planners or researchers. This topic will be picked up again in Chapter 8, 'The Future – Towards a Sustainable Electricity Supply System'.

3.8.3 Dynamic Demand Control

A scheme is being investigated that uses the already existing stored energy in millions of consumer appliances and requires the installation of *dynamic demand control* (DDC). These monitor system frequencies and switch the appliance on or off, striking a compromise between the needs of the appliance and the grid. Initially fridges and freezer applications have been investigated.

Refrigerators are 'on' in all seasons, throughout the day and night and therefore are available to participate in frequency control at all times. The total energy demand on the UK grid from domestic (excluding industrial and commercial) refrigeration has been estimated as 16.7 TWh per year, which amounts to an average load of 1.9 GW. The refrigeration load is dependent on ambient temperature, winter load being approximately two-thirds of that in summer. Daytime load is also slightly higher than that in the night time. Refrigerators are designed to handle considerable switching as they typically have a switching cycle of the order of 5 min to 1 h, depending on characteristics and contents. Any additional switching caused by frequency control should not therefore present a problem.

As an example, the power system model of Section 3.7.2 was used with the addition of DDC controlled refrigerators. For this purpose, a refrigerator with a dynamic demand controller was modelled. The aggregation of 24.9 million such appliances (one per UK household) with statistically uncorrelated behaviour and equivalent to 1320 MW of deferrable load was investigated for its response to fluctuating wind power with a reserve of 2000 MW. Figure 3.26 shows that the system with the DDC (black trace) considerably reduced the variation in frequency even though the system was operating with substantially less reserve. This is because the simulated controllers reacted more quickly than the generator governor to changes in frequency.

The graph also shows that the system frequency with and without DDC fell below the operational limit of 49.8 Hz. However, the system with DDC provided a considerable breathing time. Frequency did not fall below the operational limit until nearly 2 h after the non-DDC system. A team of engineers operating the power network would therefore be given a wider choice of generation with which to balance the system including slower-acting (and therefore possibly cheaper and more efficient) options. Also, the delay may allow generation to be scheduled more cost effectively through the electricity market that may operate a 'gate closure' (see Chapter 7) time of half an hour in advance of any particular generation slot.

It may be concluded that an aggregation of a large number of dynamically controlled loads has the potential of providing added frequency stability and smoothing to power networks, both at times of sudden increase in demand (or loss of generation) and during times of fluctuating wind or other renewable power. The devices, if incorporated on a real system, could displace some reserve and result in a significant reduction in governor activity of the remaining generators with assigned headroom. The amount of reserve displaced will depend on the extent to which low-magnitude but long-term frequency excursions can be tolerated, and on the amount of slower-acting back-up generation available, but it could be of the order of the total amount of dynamically controlled loads connected.

The potential demand that could be operated under dynamic control is considerable. Deep freeze units, industrial and commercial refrigeration, air conditioning as well as water heating systems could provide DDC. In principle, the potential available in a developed country would be several GW. The future will tell whether this potentially useful, but not essential, companion to variable renewable energy sources will attract the attention it deserves.

Worked Example 3.1

A small power system consists of two diesel generators and a wind turbine feeding a bus that supplies the total consumer demand.

The specifications of the generators are as follows.

Generator A rated at 2 MW and fitted with an isochronous governor set at 50.5 Hz. Generator B rated at 2 MW and fitted with a governor having 4% droop and a set point of 52.0 Hz. A wind turbine rated at 2 MW with variable-speed generator interfaced to the bus through a converter with no frequency control duties.

i) When the consumer demand is 3 MW and the wind turbine is supplying no power, determine the system frequency and the manner in which the demand is shared between the generators.
ii) When the demand is 3 MW and the wind turbine is supplying its rated output, determine the system frequency and the power supplied by each of the generators.
iii) For case (ii), calculate the new set point of generator B that would be required to have to maintain the system frequency at 50.5 Hz.

Model Answer

Figure 3.27 illustrates the situation and follows the principles of Figure 3.5. Plant A has an isochronous governor and is thus represented by a horizontal line. The gradient of lines B and B' reflect the 4% droop of plant B.

i) From Figure 3.27 the power from plant B, P_B, is given by the intercept of lines A and B since both plant must be operating at the same frequency. Algebraically, this is given by: $P_B/(52 - 50.5) = 2/(52 - 50)$ so $P_B = (2 \times 1.5)/2 = 1.5$ MW. There is no wind power so $P_A = 1.5$ MW also. Clearly the frequency is 50.5 Hz. Note that 3 MW is the only possible output if plant A and B are both generating.
ii) The net demand is now $3 - 2 = 1$ MW. This could either be supplied by plant A alone (system frequency 50.5 Hz) or plant B alone. In the latter case, $P_B = 1$ MW and the system frequency is 51 Hz. With generator's B set point of 52 Hz and a system

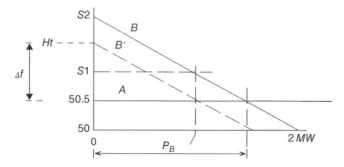

Figure 3.27 Plant governor characteristics for worked example 3.1.

frequency of 51 Hz, the governor of Generator A will unsuccessfully endeavour to lower the frequency to 50.5 by reducing its load to zero.

iii) For plant B to supply the 1 MW net load at 50.5 Hz, it must follow line B. $1/\Delta f = 2/2$ so $\Delta f = 1$ and the revised set point is 51.5 Hz.

Worked Example 3.2

Three power systems A, B and C are interconnected by tie-lines AB, BC and CA, as shown in Figure 3.28. From tests on the three systems it is found that the power–frequency constants KA, KB and KC are 650, 500 and 450 MW Hz^{-1}, respectively.

Lines AB and BC are protected so that after a time delay, they trip out of service when the power transfer exceeds 300 and 120 MW, respectively. At a time when the system frequency is 50 Hz, power system A is exporting 250 MW to B and 100 MW to C, and C is exporting 75 MW to B, a fault takes place and line AC is tripped out of service.

Calculate the consequent frequencies attained by the separate systems and the power flows over the remaining lines.

Model Answer

From Eq. (3.1):

$$\Delta P + K\Delta f = 0$$

For two coupled systems 1 and 2

$$\Delta P_{total} = \Delta P_1 + \Delta P_2 = -K_1\Delta f - K_2\Delta f = -(K_1 + K_2)\,\Delta f$$

After the tripping of line AC, system frequency is unaltered but 100 MW is now transferred to C through AB and BC. Therefore, line AB carries 350 MW that exceeds its limit and so trips out of service. As a result, the system is now split up into two separate systems: A and B + C as shown in Figure 3.28b.

For system A

$$-350 = -K_A(f_A - f) = -650(f_A - 50)$$

and

$$f_A = 50 + 350/650 = 50.54\,\text{Hz}$$

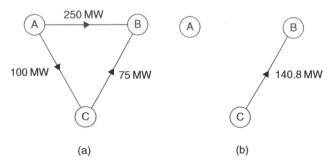

(a)   (b)

Figure 3.28 Connections between the three power systems (a) before the trip and (b), after the trip.

For system B + C

$$\Delta P_{BC} = 350 = -(K_B + K_C)(f_{BC} - 50) = -950(f_{BC} - 50)$$

and

$$f_{BC} = 50 - 350/950 = 50 - 0.368 = 49.63\,\text{Hz}$$

The 350 MW increase in demand is supplied as follows by systems B and C:

$$\Delta PB = -500(-0.368) = 184.2\,\text{MW}$$

$$\Delta PC = -450(-0.368) = 165.6\,\text{MW}$$

As a consequence, the power transfer from C to B increases by $165.8 - 100 = 65.8\,\text{MW}$ and line BC carries $75 + 65.8 = 140.8\,\text{MW}$ and as a result also trips. With all tie-lines lost, the frequencies of systems B and C can be calculated as follows:

$$\Delta P_B = 140.8 = -K_B(f_B - f_{BC}) = -500(f_B - 49.63)$$

so that

$$f_B = 49.63 - 140.8/500 = 49.35\,\text{Hz}$$

For system C

$$\Delta P_C = -140.8 = 450(f_C - 49.63)$$

thus

$$f_C = 141/450 + 49.63 = 49.94\,\text{Hz}$$

Note on these Worked Examples

On small electric grids, often one large generator is fitted with a governor that has no droop, that is, an isochronous governor that eliminates any frequency error through use of sufficient energy input to the prime mover of this generator. As already mentioned, only one machine can be operated isochronously so as to avoid instability. Both of these worked examples illustrate the operation of such small grids.

References

1 Barton, J., Huang, S., Infield, D. et al. (2013). The evolution of electricity demand and the role of demand side participation, in buildings and transport. *Energy Policy* 52: 85–102.
2 Laithwaite, E.R. and Freris, L.L. (1980). *Electric Energy: Its Generation, Transmission and Use*. McGraw Hill.
3 DTI. Ancillary Service Provision from Distributed Generation, DTI report, Contract No: DG/CG/00030/00/00, URN No: 04/1738, UK, 2004.
4 The National Grid. National Grid Seven Year Statement. available from https://www.nationalgrid.com/search-documents (accessed May 2019).
5 UKERC, The Costs and Impacts of Intermittency, Report, Imperial College, London, March 2006.

6 Farmer, E.D., Newman, V.G., and Ashmole, P.H. (1980). Economic and operational implications of a complex of wind-driven power generators on a power system. *IEE Proc. A* 127 (5): 289–295.

7 Sinden, G. (2005). *Wind Power and the UK Wind Resource*. Environmental Change Institute, University of Oxford.

8 Van Hulle F. Large scale integration of wind energy in the European power supply: analysis, issues and recommendations. Report by EWEA, December 2005.

9 ILEX Energy Consulting and UMIST. Quantifying the system costs of additional; renewables in 2020. A report by the DTI and UMIST, October 2002.

10 Grubb, M.J. (1991). Value of variable sources on power systems. *IEE Proc. C* 138 (2): 149–165.

11 Grubb, M.J. (1991). The integration of renewable energy sources. *Energy Policy* 19 (7): 670–688.

12 Parsons B. and Milligan M. Grid impacts of wind power: a summary of recent studies in the United States. European Wind Energy Conference, June 2003, Madrid, Spain.

13 Milborrow, D. (2003). Forecasting for scheduled delivery. *Wind Power Mon.* 19 (12): 37.

14 Milborrow D. False alarm. *IEE Power Engineering*, April/May 2005.

15 Kariniotakis G., Halliday J., Brownsword R., Marti, I., et al. 'What performance can be expected by short- term wind power prediction models depending on site characteristics? In European Wind Energy Conference, EWEC, 2004, London, UK, 2004.

16 Giebel, G., Brownsword, R., Kariniotakis, G. et al. (2011). *The State-of-the-Art in Short-Term Prediction of Wind Power A Literature Overview*, 2e, 110. Risø, Roskilde, Denmark: Project report for the Anemos.plus and SafeWind projects.

17 EirGrid and SONi, 2012 Curtailment Report; EirGrid and System Operator for Northern Ireland, 22 p. 2012.

18 Doherty R., All Island Grid Study. Workstream 2A: High level assessment of suitable generation portfolios for the All-island system in 2020, 2008.

19 Sinden, G. *The Practicalities of Developing Renewable Energy. Stand-by Capacity and Intermittency*. Submission to the Science and Technology Committee of the House of Lords. Environmental Change Institute, University of Oxford.

20 Streater C.J.M. Scenarios for Supply of 100% of UK Energy Requirements from Renewable Sources, REST MSc Thesis, Department of Electronic and Electrical Engineering, Loughborough University, UK, 2002.

21 Barret, M. (2007). Integrated systems modelling chapter. In: *Renewable Electricity and the Grid* (ed. G. Boyle), 157–180. Earthscan.

22 Planning of the grid integration of wind energy in Germany onshore and off-shore up to the year 2020. In International Conference on the Integration of Wind Power into the German Electricity Supply, Berlin, May 2005.

23 Docherty, R., Mullane, A., Nolan, G. et al. (2010). An assessment of the impact of wind generation on system frequency control. *IEEE Trans. Power Syst.* 25 (1): 452–460.

24 Ruttledge, L., Miller, N.W., O'Sullivan, J. et al. (2012). Frequency response of power systems with variable speed wind turbines. *IEEE Trans. Sustainable Energy* 3: 683–691.

25 Wu, L. and Infield, D. (2013). Towards an assessment of power system frequency support from wind plant – modelling aggregate inertial response. *IEEE Trans. Power Syst.* 28: 2283–2291.

26 Wu, L. and Infield, D. (2014). Power system frequency management challenges – a new approach to assessing the potential of wind capacity to aid system frequency stability. *IET Renewable Power Gener.* 8 (7): 733–739.

27 Wu L., Provision of Power System Frequency Response in the Context of High Wind Penetration, PhD Thesis, 2014, University of Strathclyde.

28 Short, J.A., Infield, D.G., and Freris, L.L. (2007, 22(3)). Stabilization of grid frequency through dynamic demand control. *IEEE Trans. Power Syst.*: 1284–1293.

29 Kundur, P. *Power System Stability and Control.* Maidenhead: McGraw Hill.

30 Tilley A.R. and Bull R.N., The Zebra Electric Vehicle Battery - Recent Advances, In Proceedings of Autotech '97 Conference, NEC, Birmingham, UK, 4–6 November 1997.

31 Etacheri, V., Marom, R., Elazari, R. et al. (2011). Challenges in the development of advanced li-ion batteries: a review. *Energy Environ. Sci.* 4: 3243–3262.

32 Carnegie, R., Gotham, D., Nderitu, D., and Preckel, P. (2013). *Utility Scale Electricity Storage Systems – Benefits, Applications and Technologies.* Internal report form State Utility Forecasting Group, University of Purdue.

33 Strategic Assessment of the Role and Value of Energy Storage Systems in the UK Low Carbon Energy Future; Report for the Carbon Trust by the Energy Futures Laboratory, Imperial College, 2012.

34 Electricity Energy Storage Technology Options, A White Paper Primer on Applications, Costs, and Benefits; EPRI report 1020676, 2010.

35 IME. Energy Storage: The missing link in the UK's energy commitments. Institute of Mechanical Engineers, 2014.

36 International Energy Agency, Technology Roadmap: Energy Storage, IEA, 2014.

37 SKM (2011). *Scottish Generation Scenarios and Power Flows: An Analysis, Report.* Sinclair Knight Merz.

38 Milborrow, D. (2007). Wind power on the grid. In: *Renewable electricity and the grid – the challenge of variability* (ed. G. Boyle). Earthscan.

39 Sinden, G.E., 2007, D Phil Thesis Environmental Change Institute, Oxford University Centre.

Further Reading

Barton, J.P. and Infield, D.G. (2004). Energy storage and its use with intermittent renewable energy. *IEEE Trans. Energy Convers.* 19 (2).

Hartnell, G. (2000). Wind on the system – grid integration of wind power. *Renewable Energy World* 3 (2).

Holt J.S., Milborrow, D.J. and Thorpe, A. Assessment of the impact of wind energy on the CEGB system. CEC Brussels, Contract No EN3W-0058-UK, 1990.

Infield, D. and Hill, J. (2015). *Literature Review: Electrical Energy Storage for Scotland.* Edinburgh: ClimateXChange (available at www.climatexchange.org.uk/reducing-emissions/energy-storage-scotland).

Patterson, W. (1999). *Transforming Electricity*. Earthscan. This book gives a readable, informative and entertaining account of the development of power systems over the past century.

SDC, Wind power in the UK, Report by the Sustainable Development Commission, UK, May 2005.

4

Electrical Power Generation and Conditioning

4.1   The Conversion of Renewable Energy into Electrical Form

Renewable energy (RE) is available in a wide variety of forms. Uniquely among
the renewable sources, biomass is available in the form of combustible fuel and thus
can play a role similar to conventional fossil fuels in electricity generation. To generate
electricity, all other renewables involve a number of conversion stages that differ from
those found in systems based on traditional fuels.

- Wind energy is available in the form of kinetic energy. The function of a wind turbine
  is to extract kinetic energy from the intercepted wind by slowing it down, and to convert this energy into rotational mechanical form that suits an electric generator. To
  improve the efficiency of conversion and for other operational reasons, the generator
  may be interfaced to the power network through a power electronic converter.
- Water power in the form of tides (potential energy) or water flow (kinetic energy)
  requires a turbine to transform this energy into rotational form for further conversion
  into electricity, again through a generator.
- The kinetic energy of water currents caused by tidal effects can alternatively be captured through an underwater turbine that can apply similar technology to that used
  for wind turbines.
- Wave energy's conversion requires specially designed devices that transform the low
  frequency energy of the waves into (most commonly) pressure energy in oil, which
  in turn is used to drive an electrical generator. Other wave energy concepts utilise
  the rise and fall of waves to drive air through a turbine coupled to an electrical generator.[1]
- Solar energy is available as radiation ranging from ultra violet to infra-red. Conversion into electricity can be implemented thermally by solar furnaces that use concentrated solar radiation to raise steam that can drive conventional turbines; this
  is known as solar thermal power generation. Alternatively, solar radiation can be
  captured and converted to electricity through solid state photovoltaic (PV) devices
  that utilise the radiation to separate charges at semiconductor junctions. The efficient operation of PV based systems depends on interfacing the PV array to the grid
  through a power electronic converter.

---

1 Note that it is common these days to refer to wave and tidal energy collectively as marine energy.

Renewable Energy in Power Systems, Second Edition. David Infield and Leon Freris.
© 2020 John Wiley & Sons Ltd. Published 2020 by John Wiley & Sons Ltd.
Companion website: www.wiley.com/go/infield-renewable-energy

This summary indicates that all renewables, except for PV systems, rely on electrome-chanical generators, also known as electrical machines, for the final stage of conversion from mechanical into electrical energy. This chapter introduces the principles of oper-ation of two classes of electrical generators the 'synchronous' and the 'asynchronous' types, both used extensively in renewable energy (RE) applications. Additionally, this chapter deals briefly with the principle of operation of the transformer, a ubiquitous device in multi-voltage level power systems. Understanding the operation of the trans-former is a necessary prerequisite for the study of the 'asynchronous' type of generator.

Power electronics play a vital role in PV and an increasingly important role in the wind power and marine energy areas. A review of power electronic devices and the converters based on them is covered in the penultimate section. The chapter concludes with a description of how electromechanical and/or power electronic converters are used in PV, wind and marine energy systems.

4.2 The Synchronous Generator

4.2.1 Construction and Mode of Operation

In an electrical generator, input mechanical power is converted into electrical output power. To get an appreciation of how this energy conversion process is carried out it is useful to look briefly at the underpinning physics. Faraday's law of electromagnetic induction [1] states that a conductor of length, l (m) moving with a velocity u (m/s) at right angles through a magnetic field of uniform flux density B (Tesla) will induce a voltage across it given by

$$v = Blu \ (volt) \tag{4.1}$$

Nature is such that this mechanical (u) to electrical (v) conversion process described by Eq. (4.1) is 'mediated' through the presence of a magnetic field (B). The equation shows that to generate a high and therefore useful voltage it is necessary to have a high magnetic flux density, a long conductor length and as high a conductor velocity as possible. All these requirements are particularly well satisfied if the mechanical-magnetic-electrical structure of the generator were arranged to be a rotating rather than a linear one. In practice, it has been found that it is preferable to have the conductors stationary and move the source of the magnetic field. With a stationary set of conductors, the problems of insulation and electrical connections are eased and centrifugal forces on the main windings are absent.

Figure 4.1a shows in outline an AC generator also known as *alternator* or *synchronous generator*. Here, the conductors that form a winding, known as the *stator*, are stationary and the source of magnetic flux rotates. The source of flux is a rotor with poles marked north and south that carries a *field winding* as shown in the figure. The field winding is fed or *excited* from an external DC source through sliding contacts known as *slip rings*. An additional reason why it is preferable to arrange that the main windings are stationary is that the DC power associated with the field is a small fraction of the power delivered by the stator winding. In fact, in some designs the field winding can be dis-pensed with completely and replaced by a rotor built with permanent magnets. There are both advantages and disadvantages to each type of excitation, which will be explored later.

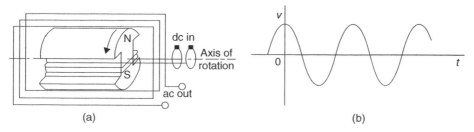

(a)   (b)

Figure 4.1 A two-pole synchronous generator. Source: Reproduced from Reference [1] with permission of John Wiley & Sons, Inc.

The flux density produced by the rotor in Figure 4.1a is maximum positive upwards along the pole axis, zero at 90° to the pole axis and maximum negative at 180°. Equation (4.1) indicates that a variable voltage with polarity reversals is generated when the field winding rotates at constant angular velocity because the magnetic field cutting each conductor increases to a maximum, decreases and successively reverses in each revolution. By proper spatial distribution of the stator winding turns and shaping of the pole faces, the generated voltage across the stator terminals in Figure 4.1a can be made to approach the sinusoid waveform shown in Figure 4.1b. It should be evident that a full rotation of the rotor will result in a full complete cycle of the sinusoidal waveform. Hence the frequency of the generated voltage in Hz (cycles per second) is identical to the angular velocity of the rotor in revolutions per second.

Appendix A explains that three-phase AC systems are universally used for the generation, transmission and utilisation of electrical energy. One of the reasons is that synchronous generators are particularly well suited for the generation of three-phase voltages. When wound for three phases, alternators make optimal use of the iron that carries the magnetic flux and of copper that carries the electric current. Figure 4.2a shows three separate one-turn windings; parts a, b and c indicate the beginnings and a′, b′ and c′ the ends of these windings. The winding axes are shifted in space with respect to each other by 120°. As a consequence, the voltages generated by the rotating

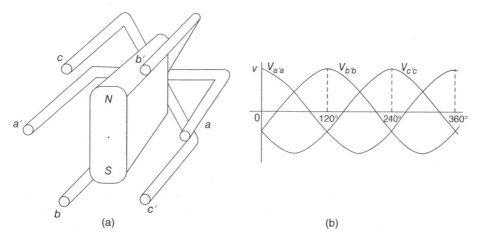

(a)   (b)

Figure 4.2 A primitive three-phase synchronous generator. (a) three separate one-turn windings and (b) the three-phase system of the AC supply.

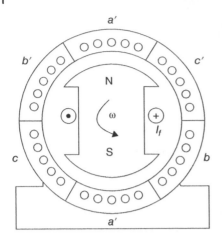

Figure 4.3 A practical three-phase synchronous generator.

field are also shifted in the time domain by one-third of a period, thus forming the three-phase system of the AC supply shown in Figure 4.2b.

A more effective arrangement for superior power output from a three-phase stator winding is shown in Figure 4.3. Here the winding is embedded and distributed in slots in an iron cylinder. The magnetic flux ϕ and therefore the flux density B generated by the rotating electromagnet is approximately proportional to the magnetomotive force (mmf) F (ampere turns) given by

$$F = N \cdot I_f \tag{4.2}$$

where N is the number of turns of the field winding (shown as having one turn in the figure) and I_f is the field or excitation current. In a practical machine, the air gap between the rotor pole faces and the internal surface of the cylinder is made small, and the cylinder itself is made of a ferromagnetic material so that minimum resistance is offered to the flow of magnetic flux. A large radial magnetic flux can therefore be produced from a moderate current I_f.

The peripheral velocity with which the conductors in the embedded windings cut the magnetic flux is, of course, proportional to the angular velocity, ω. Using Eqs. (4.1) and (4.2), the induced voltage, V, generated in a stator winding for a rotating generator is therefore given approximately by:

$$V = \mathrm{k}.\omega.I_f \tag{4.3}$$

where k is a constant of proportionality that depends on the number of turns in each winding, the distribution of the conductors in the slots, the length of the air gap and the general geometry and magnetic properties of the iron that carries the magnetic flux. Electrical machine designers take all these factors into account when designing synchronous generators so that the mechanical input power is converted into electrical power with maximum efficiency and with minimum cost of materials.

Up to this point, the alternator has been considered in stand-alone mode. The rotor driven by a prime mover induces in the stator windings a set of balanced three-phase voltages as shown in Figure 4.2b. If a balanced three-phase load is connected across the windings, a balanced set of currents will transfer power from the prime mover to the load. This stand-alone operation is the exception rather than the rule in power generation. Rather, the alternator is most likely to be required to inject power into a grid that

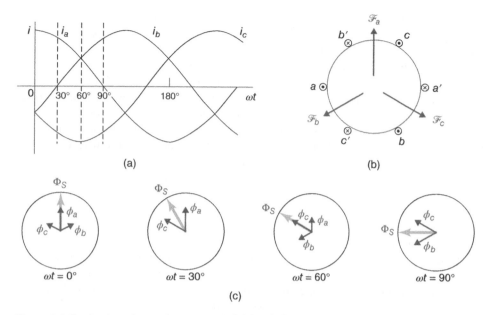

Figure 4.4 Production of a rotating magnetic field. (a) balanced currents in the three windings, (b) alternative representation of the three currents and (c) flux contributions. Source: Reproduced from Reference [1] with permission of John Wiley & Sons, Inc.

to all intents and purposes may be considered an *infinite bus*. This mode of operation is discussed in the following sections.

4.2.2 The Rotating Magnetic Field

An examination will first be made of what happens when the primitive stator windings of Figure 4.2a are connected to a three-phase supply and the rotor is absent. Figure 4.4a shows the balanced currents that will be drawn by the three primitive windings that are redrawn in Figure 4.4b. When stator winding aa′ carries the sinusoidal current i_a, Eq. (4.2) makes clear that i_a will generate a sinusoidally *pulsating* space vector field $\mathscr{F}_a$ and therefore a flux ϕ_a along the axis of coil aa′ in Figure 4.4b.

Similarly, the remaining two currents i_b and i_c in Figure 4.4a will generate pulsating fields and fluxes $\mathscr{F}_b$, ϕ_b and $\mathscr{F}_c$, ϕ_c along the axes of coils bb′ and cc′, respectively. Positive currents in the windings flowing into the $\otimes$ conductor and returning through the $\odot$ conductor in each phase produce mmfs $\mathscr{F}_a$, $\mathscr{F}_b$ and $\mathscr{F}_c$ in the directions indicated in (b). At the instant $\omega t = 0°$ in Figure 4.4a, i_a is a positive maximum and i_b and i_c are negative and one-half maximum. At this instant, the flux contributions in space can now be added as shown in Figure 4.4c for $\omega t = 0^0$. The stator flux Φ_s is the space vector sum of the three flux contributions. At $\omega t = 30^0$, the relative magnitudes of the three currents have changed and the position of the resulting stator flux has shifted anticlockwise by 30°. At $\omega t = 60°$, the current magnitudes have changed again and the resultant stator flux vector has shifted another 30°. Over a complete cycle of mains frequency, the vector Φ_s would have completed one revolution. It follows that for a 50 Hz supply the flux vector will complete one revolution in 1/50th of a second; that is, it will complete 50 revolutions per second. It is said that a rotating magnetic field (RMF) has been created by virtue of the space distribution of the three windings and of the time distribution of the currents in the three windings. This is a very important concept on which the operation of three-phase generators (and motors) depends.

To summarise, if a stationary observer were to position him- or herself inside the stator cylindrical space of Figure 4.3, he or she will observe a north-south pole pair rotating at $50 \times 60 = 3000$ rpm for a 50 Hz supply and $60 \times 60 = 3600$ rpm for a 60 Hz mains. These are known as the synchronous speeds off the 50 and 60 Hz systems for a one-*pole pair* winding. Stator windings can be arranged so that not just one pair but several pairs of poles can be created in the interior of the stator cavity. For example, a two-pole-pair arrangement will have two separate windings for each phase, each winding side occupying 30 rather than the 60° shown in Figure 4.3. It can be shown that with such an arrangement in 1/50th of a second the RMF in a 50 Hz system will advance by 180° rather than 360°. The magnetic field produced by an alternator wound for one and two-pole pairs is shown in Figure 4.5a,b, respectively.

This property is used by designers of electrical machines to generate RMF s that rotate at submultiples of the figures shown previously. The reason why such machines are useful will be explained in the following section. The general relationship linking the synchronous speed in radians per second ωS to frequency, f, and number of pole pairs, p, is:

$$\omega S = \frac{2\pi f}{p} \tag{4.4}$$

Table 4.1 shows the relationship between the synchronous speed in rpm and the pole pairs 'p' for 50 and 60 Hz systems.

4.2.3 Synchronous Generator Operation When Grid Connected

Having established the presence of a RMF within the cylindrical interior of the stator, the rotor can now be incorporated. Figure 4.6 shows the salient rotor configurations for one and two-pole pair wound stators corresponding to the flux patterns in Figure 4.5.

A one-pole pair rotor will now be considered. The rotor field current is switched on so that a north–south pole pair is created. With the rotor stationary, the interaction

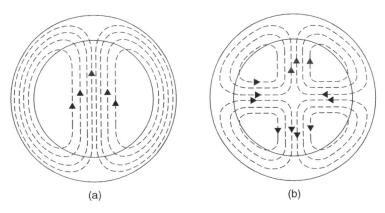

    (a)   (b)

Figure 4.5 (a) One and (b) two-pole pair wound stators.

Table 4.1 Relationship between pole pairs and synchronous speed in revolutions per minute.

p	rpm (50 Hz)	rpm (60 Hz)
1	3000	3600
2	1500	1800
3	1000	1200
4	750	900
...	...	...
...	...	...
200	15	18

Figure 4.6 Rotor configurations for a one and two-pole pair wound stators.

between the RMF and the rotor field is to say the least unproductive. As the RMF sweeps by at 50 revolutions per second or 20 ms per revolution, the only effect experienced by the rotor body is a pulsating torque as the RMF pole pair approaches and then overtakes the rotor pole pair. Due to the rotor inertia it is just not possible for the rotor to accelerate and lock onto the RMF within the required milliseconds.

Consider now that some external torque is applied to drive the rotor at increasing speed until the synchronous speed is very nearly reached. If the speed difference is close enough and decreasing, at some point the rotor north–south pair will lock onto the RMF south–north pair and the two magnet systems will rotate in synchronism. With the external torque now removed and assuming a lossless environment with no friction and windage, the rotor will rotate at synchronous speed with no requirement that any power is fed to it either mechanically from the shaft or electrically from the power network to which the synchronous machine is connected. The system is *idling* and can be thought as being suspended between motoring and generating.

In the synchronous machine, the speed of the locked magnet pair is fixed because the infinite bus frequency and therefore the RMF speed is fixed. However, the relative angular disposition of the magnetic axes of the two locked pairs is not fixed and is the mechanism that regulates the direction of energy conversion. If an external braking torque Q_t is applied to the shaft the rotor will keep rotating at the synchronous angular speed ω_s, but its magnetic axis will fall back or spatially lag the magnetic axis of the RMF. The tangential magnetic tension forces caused by the misalignment of the magnet pair is at heart the mechanism of electromechanical energy conversion. The mechanical power P_m given by

$$P_m = Q_t \, \omega_s \tag{4.5}$$

is extracted from the shaft, therefore an equal amount of electrical power must be supplied from the electrical system to which the machine is connected if the conservation of energy principle is to be satisfied. The synchronous machine is now motoring.

If the external torque is accelerating rather than decelerating the magnetic axis of the rotor is advanced with respect to the RMF axis, mechanical power is now supplied to the shaft and the energy conservation principle demands that an equal amount of power is fed into the electrical grid system. The synchronous machine is now generating.

In renewable energy applications, a high number of pole pairs may be selected for the generator as this requires a low rotational shaft speed to generate mains frequency voltages. Hydro-turbines and wind turbines are in this class. The multi-pole arrangement is particularly desirable when a wind turbine is coupled directly to a synchronous machine without an intervening gear box. In such cases, if a frequency of 50 Hz is to be generated from a wind turbine rotating at 15 rpm, Table 4.1 indicates that 200 pole pairs would be necessary!

This power generation process will now be examined from the perspective of the electrical network to which the generator is connected.

4.2.4 The Synchronous Generator Equivalent Circuit

To analyse the power flows in electrical systems, component representations are required that can be incorporated into network or circuit models. To use the available circuit analysis tools described in Chapter 5, it is necessary to build up these representations from basic circuit elements, namely: resistors, inductors, capacitors and voltage or current sources (Appendix A). Electrical power engineers over the years have developed a range of what are known as *equivalent circuits* for network simulation of electrical generators.

Here, for approximate steady-state analysis of power flows a description is required of a synchronous generator using the simplest possible equivalent circuit. The Thevenin principle explained in Appendix A can be used for example to describe the behaviour of a DC battery, by a source voltage in series with a resistance. Amazingly, this principle can also be effectively used to describe through a simple circuit, and to a good approximation, the behaviour of a generator as complex as an alternator. The steps in this transformation are not given here but can be found in any book on electrical machines (e.g. [2]). In what follows, it is assumed that the reader is familiar with the use of phasors to represent AC quantities. Readers not familiar with this concept should, at this stage study the material in the Appendix.

In Figure 4.7 the electrical generator has been reduced into a single-phase (the relationship with three-phase is dealt with later) Thevenin equivalent circuit consisting of a voltage source $V_A = V_A \angle \delta$ (the generated or 'internal' voltage of Eq. (4.3)) and a source impedance X_S, known as the *synchronous reactance*. The synchronous reactance represents in one lumped element all the internal complex interactions between the rotor and stator magnetic fields that are not of concern here. To maximise conversion efficiency synchronous machines are designed to have as low winding resistance as possible and so the source resistance representing the ohmic value of the stator winding is omitted here with little loss in accuracy.

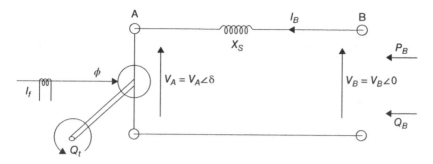

Figure 4.7 Equivalent circuit of a synchronous machine.

The equivalent circuit is shown connected to an infinite bus, that is, a network of fixed frequency f and of fixed voltage $V_B = V_B \angle 0°$ where its $0°$ angle defines it as the reference voltage.

An investigation will explore how the two available external control parameters, namely the field current I_f and the shaft torque Q_t, influence the synchronous machine and consequently the equivalent circuit behaviour. Equation (4.3) shows that $|V_A|$ depends on the field current, which is the source of the magnetic flux. It is also known that the angular disposition of the rotor magnetic axis depends on the direction and magnitude of the torque applied to the shaft. Angle δ (the *load angle*) is defined as the angle by which the axis of the rotor flux that induces V_A leads the axis of the net flux in the machine that induces V_B. This equivalence is a bridge that links the spatial disposition of rotating vector fields with the phasorial disposition of voltages in the equivalent circuit. An accelerating or 'generating' torque will result in a positive δ and in V_A leading V_B. A decelerating or 'motoring' torque will result in a negative δ and in V_A lagging V_B.

4.2.5 Power Transfer Equations

The mechanism by which power is injected into the grid is now explored. This can be done by means of the concept of complex power developed in the Appendix. The grid-connected synchronous generator of Figure 4.7 will be considered.

The complex power at end B of the line is given by

$$S_B = V_B I_B^*$$
(4.6)

Then I_B is expressed as a function of the line voltages using Kirchoff's voltage law,

$$I_B = \frac{V_B - V_A}{X_s}$$
(4.7)

Substituting Eq. (4.7) into Eq. (4.6) gives

$$S_B = V_B I_B^* = V_B \left(\frac{V_B - V_A}{X_s} \right)^* = V_B \left(\frac{V_B - V_A e^{-j\delta}}{-jX_s} \right) = j\frac{V_B^2}{X_s} - j\frac{V_A V_B}{X_s} e^{-j\delta}$$
(4.8a)

which can be written as,

$$S_B = -\frac{V_A V_B}{X_s}\sin\delta + j\left(\frac{V_B}{X_s}(V_B - V_A\cos\delta)\right) = P_B + jQ_B \tag{4.8b}$$

Hence,

$$P_B = -\frac{V_A V_B}{X_s}\sin\delta \tag{4.8c}$$

$$Q_B = \frac{V_B}{X_s}(V_B - V_A\cos\delta) \tag{4.8d}$$

If this analysis was to be carried out for terminals A in Figure 4.7, the results would be:

$$P_A = -P_B \tag{4.9a}$$

and

$$Q_A = \frac{V_A}{X_s}(V_A - V_B\cos\delta) \tag{4.9b}$$

Equation (4.9a) confirms the trivial fact that as the system is lossless, power P_B *coming out* of terminals B is equal to power P_A *fed into* terminals A.

The scalar Eqs. (4.8) and (4.9) are important in power systems technology as they describe the flow of active and reactive power of grid-connected synchronous generators.

4.2.6 Three-Phase Equations

The synchronous machine equations were derived without any reference to its three-phase nature. Assuming that the voltages in Eqs. (4.8) and (4.9) are phase to neutral in volts, the equations will give the single-phase active and reactive powers in watts and VARs, respectively. If the voltages are in kV then the active and reactive powers (both functions of voltage squared) will be in MW and MVAR.

In a balanced three-phase system, the three-phase P and Q will be three times the per-phase P and Q. Applying this to Eq. (4.8) gives

$$P_{3\varphi} = \frac{3V_A V_B}{X_s}\sin\delta = \frac{\sqrt{3}V_A \sqrt{3}V_B}{X_s}\sin\delta = \frac{V_{Al} V_{Bl}}{X_s}\sin\delta$$

where V_{Al} and V_{Bl} are line voltages.

Worked Example 4.1 *A synchronous generator of 10 ohms synchronous reactance is supplying 5 MW and 2 MVAR to an 11 kV network. Calculate the generator internal voltage.*

Rearranging Eqs. (4.8c) and (4.8d) gives

$$V_A\sin\delta = -\frac{XP_B}{V_B}$$

and

$$V_A\cos\delta = V_B - \frac{XQ_B}{V_B}$$

Therefore,

$$\tan \delta = \frac{XP}{V_B^2 - XQ}$$

In this case $P = -5$, $Q = -2$, $V_B = 11$, $X = 10$.
Hence
$\delta = 19.52°$, *and* $V_A = 13.6\,\text{kV}$

4.2.7 Four-Quadrant Operation

Consider the case where a lossless synchronous machine is run up to synchronous speed by an external prime mover. The field current is then gradually increased until the terminal voltage of the machine (the same as the internal generated voltage, as no current is taken) is made to be equal to the voltage of the local bus to which the machine is to be connected. Precise adjustments to the speed of the prime mover are made so that through some external instrumentation it is possible to detect an instant at which the internal voltage V_A exactly matches V_B in magnitude and phase. The synchronous machine can now be safely connected to the bus. This process is known as *synchronisation* and must be carried out each time a synchronous machine is to be connected to the power system.

Now arrange for the prime mover to apply an accelerating torque. This will result in a positive load angle δ and according to Eq. (4.8c) a negative active power; that is, active power injected into the power system. As expected, the machine is generating. With a braking torque on the shaft, that is, with the machine motoring, the load angle is negative and active power is supplied to the machine from the power system.

Returning to the idling state, consider now what happens if the field current is increased so that V_A is made larger than V_B but no external torque is applied so that δ and the active power are zero. Equation (4.8d) shows that the reactive power is negative; that is, the synchronous machine injects reactive power into the system, so it acts as a generator of reactive power. In this state, the machine is said to be *overexcited*.

Conversely, if the excitation current is decreased so that $V_A < V_B$, the reactive power flow is reversed, the machine becomes a consumer of reactive power and it is said to be *underexcited*.

4.2.8 Power–Load Angle Characteristic

Equation (4.8c) is plotted in Figure 4.8 to illustrate the dependence of the generated active power on the load angle.

Note that a synchronous generator (SG) connected to an infinite bus is capable of generating a maximum active power, P_{Bmax}, at $\delta = 90^0$. Any additional applied mechanical torque will increase the load angle beyond 90^0 with a consequential decrease in electrical power. Physically, the peripheral magnetic forces linking the two fields together are insufficient to maintain the locking effect. The power balance between mechanical and electrical powers has now been lost, the excess mechanical power accelerates the rotor beyond synchronous speed and the synchronous generator has lost its ability to act as a stable power converter. In this unstable regime, the SG is described as having *lost synchronism*; that is, its rotor generates an internal voltage V_A of a higher frequency

Figure 4.8 Power angle characteristic of a SG.

than the infinite bus voltage V_B. Operation under this condition, known as *pole slipping*, results in large overcurrents, is highly undesirable and protection equipment will be brought into action to disconnect the synchronous machine from the network.

During system contingencies, violent transient changes may take place to the prime mover or in the power system that cause swings in the load angles of synchronous generators. To ensure safe stability margins so that pole slipping is prevented, the steady-state load angles of SGs are kept well below 30°. Power system stability is a major topic in its own right involving the simultaneous solution of the differential equations characterising all the network components. Although this is beyond the scope of this book, it is worth some comments on system stability since the connection of large amounts of RE sources onto the grid will alter the system dynamics. These can be found in Section 5.9.

4.3 The Transformer

4.3.1 Transformer Basics

The transformer is an indispensable part of any power system operating at a range of voltages. The transformer mode of operation is included here as it provides a useful aid to the understanding of the mode of operation of induction generators, which are used extensively in wind turbines.

The properties of an inductance are discussed in the Appendix. A current carrying inductor, usually wound in the form of a coil, generates a magnetic flux. This flux links with the turns of the coil or the *winding* and induces a voltage in these turns if the current and therefore the flux is increased or decreased. It stands to reason that if a changing magnetic field produced by one inductor were to 'link' with the turns of wire in another adjacent inductor, it would induce a voltage in the *unpowered* inductor. This phenomenon is called *mutual inductance* and the *transformer* is a device constructed to exploit this effect.

By having two inductors coupled together by a common magnetic field path, it is possible to transfer energy from one inductor circuit to the other. In order for this to work, the magnetic field has to be constantly changing in strength, otherwise no voltage will be induced in the unpowered winding. Thus, the transformer is essentially an AC device.

Figure 4.9 The transformer. Source: Reproduced from Reference [1] with permission of John Wiley & Sons, Inc.

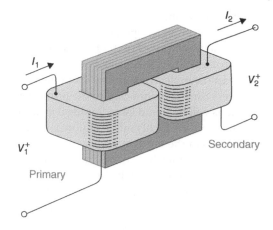

The powered winding of a transformer is called the *primary*, while the unpowered winding is called the *secondary*.

Figure 4.9 shows a transformer in outline. The primary winding labelled 1 is connected to an AC supply V_1. The winding is wound round a substantial closed ferromagnetic core. This provides a very low 'magnetic resistance' to the flow of magnetic flux in comparison to the surrounding air. Because of the very low resistance to the flux flow, only a very low current, known as the *magnetising current,* is required from the supply to set up an mmf that circulates a flux, which induces a voltage in the primary that exactly balances V_1. For the present the effect of this low current may be disregarded. Because of the low 'magnetic resistance' of the ferromagnetic core it can be assumed that practically all the generated magnetic flux is constrained to flow within the core, with only a small amount of *leakage flux* taking paths in the surrounding air, which will also be disregarded at this point. The consequence of this is that the secondary winding, which is also wound round the transformer core, links exactly the same flux as the primary winding. Finally, if it is assumed that the windings consisted of large cross section copper wire their ohmic resistance could be disregarded. All these simplifications lead to the concept of the *ideal transformer*.

In an ideal transformer, the voltage induced in an inductor of N turns linking a flux varying at $d\phi/dt$ is proportional to N. Hence for the transformer in Figure 4.9, it can be written that

$$\frac{V_1}{V_2} = \frac{N_1}{N_2} \tag{4.10}$$

Suppose that the secondary is connected to a resistor that draws current I_2 with the consequence that power $V_2 I_2$ is extracted from the secondary. The energy conservation principle requires that this power is supplied from the source to which the primary is connected. It follows that a primary current I_1 is established the value of which can be determined from

$$V_1 I_1 = V_2 I_2 \tag{4.11}$$

and from Eq. (4.10) this gives

$$\frac{I_1}{I_2} = \frac{N_2}{N_1} \tag{4.12}$$

Note that an open circuit or a short circuit on the secondary winding of an ideal transformer are seen as an open or short circuit, respectively, on the primary side.

An ideal transformer with identical primary and secondary windings would manifest equal voltage and current in both sets of windings. In a perfect world, transformers would transfer electrical power from primary to secondary as efficiently as though the load were directly connected to the primary power source, with no transformer there at all, but as will be found later this ideal goal cannot be realised in practice. Nevertheless, transformers are highly efficient power transfer devices with no moving parts achieving efficiencies in the upper nineties.

The transformer has made long-distance transmission of electric power a practical reality, as AC voltage can be 'stepped up' and current 'stepped down' for reduced ohmic resistance losses along power lines connecting generating stations with loads. A transformer that increases voltage from primary to secondary (more secondary winding turns than primary winding turns) is called a *step-up* transformer. Conversely, a transformer designed to do just the opposite is called a *step-down* transformer.

4.3.2 The Transformer Equivalent Circuit

The next task is to develop an equivalent circuit capable of representing realistically the behaviour of a transformer in studies aimed at determining the flow of power in electrical systems. Figure 4.10 shows how such an equivalent circuit can be built up starting with the ideal transformer in the hatched box. The small but finite current required to set up the flux in the core is simulated by the presence of the shunt inductance L_m, the *magnetising inductance*, that has a large value in henries.

Next, the finite resistance of the two transformer windings can be simulated by the series resistors R_1 and R_2. To keep losses low, these have low numerical values. Finally, any magnetic flux not contained in the core is free to store and release energy rather than transfer it from one coil to the other. Any energy thus stored by this *uncoupled* flux manifests itself as an inductance in series with the relevant winding. This stray inductance is called *leakage inductance* and is represented in the equivalent circuit by L_1 and L_2 both having numerical values much smaller than L_m.

The equivalent circuit of Figure 4.10 can be further simplified if the R_2 and L_2 are transferred to the primary so that the primary and secondary resistances and inductances could be lumped into just two series components. This can be done through the conservation of energy principle. The transferred resistance from secondary (2) to primary (1),

Figure 4.10 The transformer equivalent circuit.

Figure 4.11 Simplified transformer equivalent circuit.

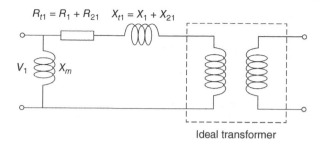

$R_{t1} = R_1 + R_{21}$ $X_{t1} = X_1 + X_{21}$

Ideal transformer

which can be called R_{21}, should have such a value that when it carries I_1 should dissipate the same amount of power as R_2 when carrying I_2; that is, $R_2 I_2^2 = R_{21} I_1^2$. Substitution of the number of turns for currents from Eq. (4.12) gives:

$$R_{21} = R_2 \left(\frac{N_1}{N_2} \right)^2 \tag{4.13}$$

Similar logic based on the reactive power conservation principle (Appendix A) can be applied for the transfer of the secondary winding reactance to the primary:

$$X_{21} = X_2 \left(\frac{N_1}{N_2} \right)^2 \tag{4.14}$$

A new equivalent circuit incorporating these changes is shown in Figure 4.11. Here R_{t1} and X_{t1} are the total winding resistance and reactance, respectively, referred to the primary winding. In this circuit the magnetising reactance X_m is connected across the mains with little loss in accuracy because it can be shown that the voltage drops across R_{t1} and X_{t1} are small. This equivalent circuit is frequently used to calculate the effect of a transformer in a power network and will be used later to develop an equivalent circuit for the asynchronous generator.

4.3.3 Further Details on Transformers

The manufacturer of an electrical machine such as a transformer will indicate on the nameplate the normal operating conditions, for example: '11,000:415 V, 50Hz, 500kVA' The *rated output* of 500 kVA can be maintained continuously without excessive heating and the consequential deterioration of the winding insulation. Because the heating is dependent on the square of the current, the output is rated in apparent power (kVA) rather than active power (kW). When supplying a zero power-factor load, a transformer can be operating at rated temperature while delivering zero active power.

On large transformers, taps on the windings allow small adjustments on the turns ratio. Often, these taps are operated by an automatic tap-changer that maintains the voltage, usually on the secondary, at a fixed value irrespective of the load on the transformer.

It can be shown [1] that the size and therefore, weight and cost of a transformer is intimately related to the frequency of operation. The higher the frequency the lower the weight and cost. For these reasons, in power electronic systems whenever a transformer is to be used, higher frequencies are employed, a topic to be revisited later

4.4 The Asynchronous Generator

4.4.1 Construction and Properties

Asynchronous or *induction machines* operating as motors are the most widely used electromechanical converters. In an induction machine the stator is identical to the one for synchronous machines as shown in Figure 4.3 in which three-phase currents supplied to the stator produce a RMF. The rotor, however, is radically different and it has neither an external magnetising source nor permanent magnets. Instead, alternating currents are injected in the rotor from the stator through induction or transformer action, hence the useful parallel with the operation of a transformer. It is the interaction between these induced rotor currents and the stator RMF that results in torque production.

In its most common form, the rotor consists of axial conductors shorted at the ends by circular rings to form a *squirrel cage* or just *cage* as shown in Figure 4.12.

Although for the purposes of renewable energy sources there is interest in the generation mode, it is easier initially to understand the operation of the induction machine from a motoring perspective.

As the stator RMF moves at ω_s (given by Eq. (4.4)) past the stationary rotor conductors, three-phase electromotive forces (EMFs) are induced in the spatially shifted rotor conductors by flux cutting action. The resulting rotor currents, according to Lenz's law [1], are of such magnitude and direction as to generate a torque that speeds up the rotor. If the rotor were to achieve speed ω_s, there would be no change in flux linkage, no induced voltage, no current in the rotor conductors and therefore no torque. For EMFs to be induced in the rotor conductors they should possess some relative speed with respect to the stator RMF. For motoring, the rotor therefore turns at a lower speed ω_r.

Figure 4.12 Induction machine with cage rotor (1: shaft, 2: cage rotor, 3: stator three-phase winding, 4: terminal box, 5: stator iron core, 6: cooling fan, 7: motor frame). Source: Reproduced with permission of Asae Brown Boveri Ltd.

It can be shown that the rotor currents themselves produce an RMF whose speed depends on the frequency of these currents. For a constant torque interaction to take place, the rotor RMF must rotate in synchronism with the stator RMF as in the case of the synchronous machine. How is this accomplished if the rotor rotates at a speed lower than ω_s?

The difference between ω_s and ω_r is expressed as a ratio with respect to ω_s and is known as *slip*, *s*, where

$$s = \frac{(\omega_s - \omega_r)}{\omega_s} \tag{4.15}$$

Therefore

$$\omega_r = (1 - s)\,\omega_s \tag{4.16}$$

The relative motion between the stator's and rotor's RMF determines how frequently the stator RMF cuts the rotating rotor conductors, so the frequency of the rotor induced voltages and currents f_r is:

$$f_r = sf \tag{4.17}$$

where f is the mains frequency. The frequency of the rotor currents determines ω_{RR}, the speed of the rotor RMF with respect to the rotor:

$$\omega_{rr} = 2\pi f_r / \text{p} = 2\pi sf / \text{p} = s\omega_s \tag{4.18}$$

The speed of the rotor RMF with respect to the stationary stator is the rotor speed *plus* the rotor RMF's speed with respect to the rotor:

$$\omega_r + \omega_{rr} = (1 - s)\,\omega_s + s\,\omega_s = \omega_s$$

It can be concluded that the rotor and stator RMFs rotate together at synchronous speed as required for a uniform torque to be developed as in the synchronous machine. However, in contrast to the synchronous machine, the rotor RMF is produced through *induction* from the stator. The larger the applied braking torque the higher the slip, the larger the induced EMFs and resulting rotor currents, and the stronger the interaction between the two RMFs to produce an electrical torque equal and opposite to the braking torque. The induction motor therefore exhibits a small decrease in speed with increments in braking torque.

An ideal, lossless induction machine could be imagined to operate at zero slip. This is equivalent to the 'idling' state of the fixed speed synchronous machine. The vital difference, however, in the induction machine is that motoring or generating torques will be accompanied by a decrease or increase of speed below or above synchronous, respectively. For generating, Eq. (4.16) now gives a negative slip. The induction machine will move seamlessly from motoring into generation mode as the external torque changes from a decelerating to an accelerating type. Indeed, in small wind turbines, it is very common to find that the induction generator was originally designed as a motor and has been employed as a generator without any modification.

Worked Example 4.2

A six-pole 50 Hz induction motor runs at 4% slip at a certain load. Calculate the synchronous speed, the rotor speed, the frequency of the rotor currents, the speed of the rotor RMF with respect to the rotor and the speed of the rotor RMF with respect to the stator.

Worked Example: Model Answer

The synchronous speed from Eq. (4.4) is: $N_s = f/p = 50/3 \text{ rps} = 50 \times 60/3 = 1000 \text{ rpm}$

The rotor speed from Eq. (4.16) is: $(1 - s)N_s = (1 - 0.04) \times 1000 = 960 \text{ rpm}$

The frequency of the rotor currents is: $f_r = sf = 0.04 \times 50 = 2 \text{ Hz}$

The speed of the rotor RMF with respect to the rotor:

$N_{rr} = f_r \times 60/p = 2 \times 60/3 = 40 \text{ rpm}$

The speed of the rotor RMF with respect to the stator:

$N_r + N_{rr} = 960 + 40 = 1000$

(i.e. the rotor and stator RMFs rotate together).

4.4.2 The Induction Machine Equivalent Circuit

The induction machine can be viewed as a transformer with a rotating secondary. Imagine an induction machine with its rotor mechanically locked; that is, at standstill. The stator RMF will be rotating at ω_S with respect to the rotor and inducing in each phase the voltage E_2 at mains frequency f. The current that flows in each phase will be

$$I_2 = \frac{E_2}{R_2 + j2\pi f L_2} = \frac{E_2}{R_2 + jX_2}. \tag{4.19}$$

where R_2 and L_2 are the effective per-phase resistance and inductance of the rotor winding and X_2 is the rotor reactance at mains frequency. At standstill the slip $s = 1$ and the rotor voltages and currents are of the stator frequency, f. At any other rotor speed, the slip is s, the induced voltage is sE_2, the rotor frequency is sf and the rotor reactance is sX_2. For the rotor current at slip s the more general expression applies:

$$I_2 = \frac{sE_2}{R_2 + jsX_2} = \frac{E_2}{(R_2/s) + jX_2} = \frac{E_2}{R_2 + R_2[(1 - s)/s] + jX_2} \tag{4.20}$$

since

$$R/s = R_2 + R_2[(1 - s)/s]$$

Equation (4.20) provides the rationale for the equivalent circuit shown in Figure 4.13. This circuit resembles that of the secondary winding of the transformer (Figure 4.10) but with a variable resistive load connected to its output. The energy conservation principle indicates that the electrical power transferred to the rotor is the real part of $E_2 I_2^*$. The power lost irreversibly in the rotor ohmic resistance is $R_2 I_2^2$ and the remainder, that is, $I_2^2 R_2[(1 - s)/s]$, must be and indeed is the electrical power converted into mechanical power.

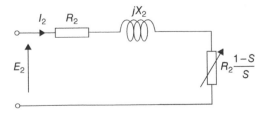

Figure 4.13 Equivalent circuit of induction machine rotor.

Figure 4.14 Induction machine stator-rotor equivalent circuit.

Figure 4.15 Induction machine equivalent circuit referred to the stator.

The transformer equivalent circuit analogy is extended in Figure 4.14 to include the stator parameters. Here, R_1 and X_1 represent the stator winding resistance and inductance while X_m represents the magnetising reactance drawing the current necessary to establish the RMF. The dashed line corresponds to the air gap interface across which energy is transferred from the stator to the rotor.

In a further simplification, the ideal transformer can be omitted by transferring elements from secondary to primary using the transformation ratio. In Figure 4.15 R_s and X_s are the stator winding resistance and reactance, respectively. The elements R_r and X_r represent the rotor resistance and reactance respectively referred to the stator using the rotor-stator transformation ratio. The product $R_r[(1-s)/s]I_s^2$ represents the electrical power per-phase converted into mechanical power. This equivalent circuit indicated that when the rotor is locked, $s = 1$, $R_r[(1-s)/s] = 0$, so all the input to the rotor is converted into heat in R_r. When $s < 1$, the energy into the rotor is partly converted into heat in the winding resistances and partly into mechanical form.

With the induction machine generating, the slip is negative and the notional resistance $R_r[(1-s)/s]$ is also negative. This is perfectly consistent with circuit analysis. A positive RI^2 implies irreversible conversion of electrical energy into thermal energy. A negative RI^2 implies the conversion of some other type of energy (in this case mechanical) into electrical.

Irrespective of whether the machine is motoring or generating into the network to which it is connected, it supplies the reactive voltamperes absorbed by all the inductive components of the equivalent circuit. This must be so that only positive or negative real power is associated with the mechanical/electrical energy conversion in the resistance $R_r(1-s)/s$. The consequence is that induction generators always absorb reactive power from the network.

4.4.3   The Induction Machine Efficiency

If the total electrical input power per phase fed into the stator is P_s, the power crossing the motor's air gap; that is, the power per phase transferred from the stator to the rotor is $P_r = P_s - R_s I_s{}^2$. All of P_r is dissipated in $\{R_r + R_r[(1-s)/s]\} = R_r/s$ so, $P_r = \left(\dfrac{R_r}{s}\right)I_r^2$

Hence, the rotor copper loss is:

$$R_r I_r{}^2 = sP_r \tag{4.21}$$

Subtracting the rotor copper loss from P_r gives the average per-phase mechanical power $P_m = P_r - R_r I_r{}^2$ that through substitution from Eq. (4.21) gives:

$$P_m = (1-s)P_r \tag{4.22}$$

The developed torque Q_m of the motor is its total mechanical Power $3P_m$ divided by the motor shaft speed ω_r, therefore

$$Q_m = \frac{3P_m}{\omega_r} \tag{4.23}$$

Substituting Eqs. (4.16) and (4.22) into Eq. (4.23) gives

$$Q_m = \frac{3(1-s)P_r}{(1-s)\omega_s} = \frac{P_{ag}}{\omega_s} \tag{4.24}$$

where $P_{ag} = 3P_r$ is the total three-phase power crossing the air gap.

Neglecting the stator copper losses and rotating mechanical losses, the efficiency of an induction motor is given approximately by:

$$\eta = \frac{P_{out}}{P_{in}} = \frac{P_m}{P_r} = \frac{(1-s)P_r}{P_r} = (1-s) \tag{4.25}$$

For an induction generator (s is negative) the power flow is in the reverse direction hence, approximately

$$\eta = \frac{P_{out}}{P_{in}} = \frac{P_r}{P_m} = \frac{1}{(1-s)} \tag{4.26}$$

Equations (4.25) and (4.26) indicate that for the conversion efficiency to be high, s at full load must be as small as possible.

Real induction generators have losses that have not been taken into account in this simplified analysis. The mechanical power available to produce electricity is reduced by windage and other mechanical frictional losses within the generator. Additionally, electrical and magnetic losses within the rotor reduce the power that is transferred from the rotor across the air gap to the stator. Finally, in the stator there are more losses associated with the winding resistance and the setting up of the magnetic excitation in the shunt branch of the equivalent circuit. As a consequence, large induction generators have efficiencies in the region of 90%. These extra losses will be referred to in a later section.

4.4.4   The Induction Machine Speed-Torque Characteristic

An important characteristic of any electromechanical converter is its speed-torque relationship. For the induction machine the developed torque Q_m from Eqs. (4.23), (4.22) and (4.16) is

$$Q_m = \frac{3P_m}{\omega_r} = \frac{3(1-s)P_r}{(1-s)\omega_s} = \frac{3P_r}{\omega_s}$$

Substituting for

$$P_r = \frac{R_r}{s}I_s^2 = \frac{R_r V_t^2}{sZ^2}$$

and using the equivalent circuit of Figure 4.15 gives

$$Q_m = 3\left(\frac{R_r}{s}\right)\frac{I_s^2}{\omega_s} = \frac{3R_r}{s\omega_s}\left[\frac{V_s^2}{(R_s + R_r/s)^2 + (X_s + X_r)^2}\right] \tag{4.27}$$

This relationship is non-linear and can be generalised to describe typical performances of an induction machine by normalising it in terms of torque and speed. Taking as normal torque the rated torque and as normal speed the synchronous speed, the normalised relationship of Eq. (4.27) for a typical induction generator is plotted in Figure 4.16 with R_r as a parameter.

Note that, as for the synchronous machine, there is a maximum or *pullout* torque beyond which the generator will accelerate uncontrollably. However, this condition is far away from the normal operating regime. The curve for rotor resistance 'R_r' represents the performance of a typical induction machine with low rotor resistance and shows that the variation in speed from zero input torque to rated torque varies by about 3 to 4%. For a small s, R_r/s is large compared to R_r and $X_s + X_r$ and to a good approximation Eq. (4.27) can be written as

$$Q_m \approx \frac{3V_s^2 s}{\omega_s R_r} \tag{4.28}$$

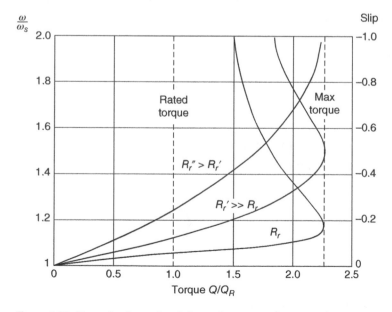

Figure 4.16 Normalised speed and slip against torque for generating.

Equation (4.28) indicates that in the normal operating range (zero to rated torque), torque is directly proportional to slip and therefore speed and inversely proportional to R_r. Curves for $R_r' > R_r$ and $R_r'' \gg R_r$ are also plotted on Figure 4.16. By selecting the value of the rotor resistance, a designer has the ability to change the slope of the torque-speed characteristic. If a substantial variation of speed with torque is required, the rotor can be designed to have a large resistance. The downside of such an arrangement is the unacceptable reduction in efficiency.

A method to access the rotor windings and therefore exploit the property of speed change is to arrange a rotor that has coils rather than short circuited bars, with the coil terminals connected to slip rings and brushes so that additional external resistance can be connected in series with the windings. In such a *rotor wound* induction machine, the rotor winding is similar to that on the stator. The disadvantages of the wound rotor induction generator include higher capital cost and higher maintenance cost.

Worked Example 4.3

A wind turbine rated at 450 kW has the following induction generator parameters in ohms: $R_1 = 0.01$, $X_1 = X_2 = 0.15$, $R_2 = 0.01$ and $X_m = 6$. At a time when it is supplying its rated output the slip is 0.01. Calculate the network voltage and the power factor at which the induction generator is supplying power to the grid using the simplified equivalent circuit.

Model Answer

The circuit in Figure 4.17 brings to mind the synchronous generator equivalent circuit of Figure 4.7. The reactance to resistance ratio of the series impedance in Figure 4.17 is $0.3/0.02 = 15$, hence, to a good approximation, Eqs. (4.8) and (4.9) can be used. The mechanical power converted from the wind into electrical power per-phase is $450\,000/3 = 150\,000$ W. This appears as a 'negative' power dissipation in R_{em} therefore the voltage V_A across R_{em} is given from:

$$150\,000 = \frac{V_A^2}{R_{em}}$$

and

$$V_A = 387 \text{ V}$$

Figure 4.17 Equivalent circuit for Worked Example 4.3.

Using Eq. (4.8c)

$$150000 = \frac{V_A V_B}{X} \sin \delta = \frac{387 V_B}{0.3} \sin \delta$$

and

$$V_B \sin \delta = 116.28$$

The reactive power at end A is zero, hence from Eq. (4.8d)

$$0 = \frac{V_A}{X}(V_A - V_B \cos \delta) = \frac{387}{0.3}(387 - V_B \cos \delta)$$

and

$$V_B \cos \delta = 387$$

hence

$$\tan \delta = \frac{116.28}{387} \qquad and \qquad \delta = 16.7°$$

$$giving \qquad V_B = 404V \qquad i.e. \qquad 404\sqrt{3} \approx 700 \qquad line\ volts$$

The reactive power associated with end B of the line is given by Eq. (4.9b)

$$Q_B = \frac{V_B}{X}(V_B - V_A \cos \delta) = \frac{404}{0.3}(404 - 387 \cos 16.7) = 44874 VAR$$

The Q absorbed by the shunt reactance $X_m = 6\ \Omega$ is given from

$$Q_m = \frac{404^2}{6} = 27,202$$

The total Q per phase is $= 44\,874 + 27\,202 = 72\,076$. Hence, the power factor is

$$\cos \tan^{-1} \frac{72076}{150000} = 0.9$$

4.4.5 Induction Generator Reactive Power

Worked Example 4.3 shows that the induction generator is a source of active power but a sink of reactive power. Even when the real power output from an induction generator is zero it will still draw considerable reactive power through X_m to magnetise its iron core (3×27.2 kVARs in the worked example). As increasing torque is applied and more real power is exported to the network, extra reactive power is absorbed due to the reactive power consumed by the series reactance (3×44.87 kVARs in the example). A typical relationship between active and reactive power for an induction generator is shown in Figure 4.18.

The induction generator power factor will vary from zero at A to around 0.9 at B. In order to improve the power factor, it is often necessary to fit local Power Factor Correction (PFC) capacitors at the generator terminals (Appendix). These have the effect of shifting the overall characteristic downwards to A' B'. The amount of reactive power 'compensation' required depends on a number of technical and economic factors.

4.4.6 Comparison Between Synchronous and Asynchronous Generators

Table 4.2 provides an overall comparison of the characteristics of induction and synchronous generators with particular reference to renewable energy applications.

Figure 4.18 Relationship between active and reactive power for an induction generator.

Table 4.2 Comparison between synchronous and asynchronous generators.

Synchronous generator	Induction generators
Features	
• Efficient	• Moderately efficient
• Expensive	• Less expensive
• Requires maintenance	• Rugged and robust, little maintenance
• Reactive power flow can be controlled through field current	• Sink of reactive power
• Fixed speed hence very stiff	• Small change in speed with torque, hence more compliant
• Responds in an oscillatory manner to sudden changes in torque	• Responds to sudden torque inputs in a non-oscillatory way
• Can be built with permanent magnets for large number of pole pairs and low rotational speed ('ring' form)	• Cannot be built economically for low rotational speeds
• Suitable for variable speed operation through a power electronic interface	• Suitable for variable speed in its rotor wound form in conjunction with a power electronic converter
• Suitable for connection to weak networks. Used in autonomous systems	• Suitable for week networks only in conjunction with power electronics
• Requires special synchronisation equipment to connect to the network	• Can be simply synchronised to the network
Use in RE generation	
• Used in wind power mainly in its 'ring' form for gearless coupling to a wind turbine	• Used extensively with a gear box in wind power
• Variable speed provided through a DC link power electronic interface	• Variable speed provided with power electronics in the rotor wound form
• Used in water power when reactive power control is required	• Used in water power with gear box

4.5   Power Electronics

4.5.1   Introduction

Power electronics is concerned with the application of electronic devices to *control* and *condition* electrical power. Power control involves the regulation of the power transfer from the renewable energy generator to the power network to either continuously maximise this transfer as the available resource changes or to limit the transfer for operational reasons. Power conditioning involves the transformation of power from one voltage/current/frequency/waveform to a different voltage/current/frequency/waveform.

A power electronic interface that controls and/or conditions power is referred to as a *converter*. Converters are interposed between the RE generator and the network and often carry the total power transfer. They should be designed so that as little power as possible is lost in this transfer and that their capital cost is as slow as possible per watt transferred. Transformers can be viewed as power conditioning devices. Converters, not unlike transformers, are rated in volt-amps (VA).

The devices (transistors etc.) used in power electronic converters are made from semiconductor silicon, like those in a computer processor, but physically they are much larger and usually they are discrete – one device per piece of silicon. A converter will typically contain between 1 and 100 power-semiconductor devices.

Power electronic converters are found in an enormous range of applications and sizes, from the 2 GW UK-France Channel DC Link down to the 40 W energy-saving light bulbs in our homes. In renewable energy, power electronic converters are already used in most PV systems and many wind turbines. It is also expected that they will be used in the future in practically all the emerging wave and tidal power technologies.

Despite the enormous range of converter sizes and applications, they are mostly based on just a handful of device types and basic circuit configurations. This section will discuss the most common of these. The aim is not that the reader will be able to design converters, but that she/he will understand their specifications, potential uses and limitations. The following section will deal with the application of these converters in the conditioning of power from renewable energy sources.

More details on solid state switching devices and converters can be found in many books on power electronics, for example, Reference [3]. Some of the most common devices are reviewed next.

4.5.2   Power-Semiconductor Devices

4.5.2.1   Diodes

The *diode* in Figure 4.19 conducts current from anode to cathode (in the direction of the arrow) whenever the anode is made positive with respect to the cathode, that is positive *bias voltage* is applied across the device, and blocks current in the reverse direction when the device is negatively biased. The diode has no capability to control the current after it has been established. Diodes are readily available with ratings up to thousands of volts and thousands of amps. They are very reliable and very cheap (per volt-amp) and very useful.

4.5.2.2   Thyristors

In contrast to the diode, the *thyristor* (Figure 4.20) has a control capability through a third electrode known as the *gate*. Even if the thyristor is positively biased, no current

Figure 4.19 Diode.

Figure 4.20 Thyristor.

will flow through the main anode-cathode circuit until a small *pulse* has been applied to the gate. Thereafter, the thyristor acts like the diode so that the current will *continue* to flow through the anode-cathode circuit until that current reduces to zero. The thyristor itself cannot reduce or switch off the current. Unlike the diode though, after the current has reached zero, the thyristor regains its capability to block current in the forward direction

Thyristors were the first commercially-available controllable power-semiconductor devices and were used in converters of all sizes. They are still the cheapest controllable devices and are used in very high-power converters (hundreds of MVA) and in smaller cost-sensitive applications.

Gate Turn-Off thyristors (GTOs) and Integrated-Gate Commutated Thyristors (IGCTs) are devises based on the thyristor principle but capable of turning themselves both on and off. These derivatives are expensive and are being superseded by advances in transistor technology.

4.5.2.3 Transistors

Bipolar Junction Transistors (BJTs), Metal Oxide Semiconductor Field Effect Transistors (MOSFETs) and Insulated Gate Bipolar Transistors (IGBTs) (Figure 4.21) are all significantly more expensive than simple thyristors, but have the big advantage that they can be turned off by a control signal.

Practical and economic considerations currently favour the use of MOSFETs for small converters up to about 30 kVA, which includes most PV inverters. IGBTs are used in converters up to about 10 MVA, which includes wind turbine applications.

Figure 4.21 Transistor types.

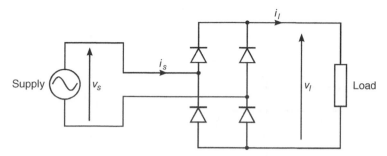

Figure 4.22 Circuit of a single-phase diode bridge rectifier.

4.5.3 Diode Bridge Rectifier

AC power is fed from the supply to the diode junctions. It is *rectified* by the four diodes, which form the *bridge*, and the resulting DC power is supplied to the load on the right (Figure 4.22). Thus, the circuit converts the AC power to DC. Assuming, for simplicity, that the load is purely resistive, the resulting waveforms are shown in Figure 4.23.

The voltage supplied to the load is DC, in that it is always positive, but it is far from smooth because it contains a considerable *ripple*. Very few practical loads would tolerate a supply like this except those, for example heating resistors, that would be equally happy on an AC supply. For this reason, it is common to add a smoothing capacitor as shown in Figure 4.24.

The waveforms in Figure 4.25 show that as before, the diode bridge is rectifying AC power to DC. But now, the voltage supplied to the load v_l is almost constant; a bigger capacitor would reduce the ripple even further. The capacitor acts as an energy store so that when the output voltage from the rectifier is lower than the voltage across the capacitor, energy is supplied to the load from the capacitor. When the output voltage from the rectifier is larger than the capacitor voltage, the capacitor is charged from the supply. This charging effect is provided by a series of pulses shown as i_b in the figure. Because the duration of the pulses is short, their amplitude must be high, if the same average

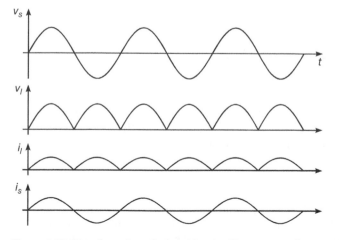

Figure 4.23 Waveforms in a diode bridge rectifier connected to a purely resistive load.

Figure 4.24 Circuit of a diode bridge rectifier with a smoothing capacitor, and a supply impedance.

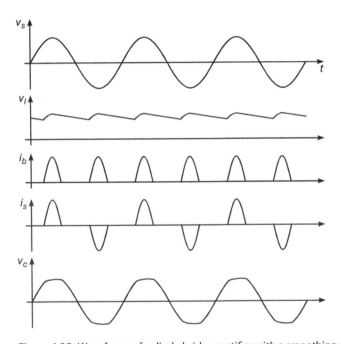

Figure 4.25 Waveforms of a diode bridge rectifier with a smoothing capacitor.

power is to be supplied to the load as it was supplied before the capacitor insertion. The downside of this arrangement is that the current drawn from the AC source i_s is no longer a sine wave. Instead, it is a series of positive and negative pulses.

4.5.4 Harmonics

The non-sinusoidal current i_s in Figure 4.25 may cause problems for the AC supply network. In particular, transformers and cables in the network will experience additional heating. Despite this, the circuit shown in Figure 4.23 and variants with the same problem are very widely used, because they are cheap. Televisions and computers are a problem in this regard, mainly because there are so many of them. Indeed, in areas where the load is dominated by televisions and computers, it is common for the customer's voltage waveform v_c in Figure 4.25 to be *flat-topped*, due to the non-sinusoidal voltage drop in the supply impedance. Plotting instantaneous supply current i_s against voltage v_c would show a highly non-linear relationship. Appliances such as televisions and computers are sometimes called non-linear loads.

Another way of describing the problem is by considering the harmonic content of the waveforms that can be obtained through Fourier analysis. For example, the supply current i_s in Figure 4.25 has a very large third harmonic component. This is a particular problem in three-phase systems (in just about every large power system in the world) because the third harmonics in the neutral conductor do not cancel (Appendix A). Thus, the normal assumption that the neutral conductor carries zero current in a balanced three-phase system no longer holds. In general, all harmonic currents cause undesirable heating in the transformers and cables of the supply system. In a bid to prevent excessive harmonic currents in distribution networks, electricity utilities have instigated the introduction of regulatory standards that set maximum permissible levels for the harmonic currents that individual appliances may cause.

In general, all power electronic converters cause some harmonic currents in the AC network to which they are connected. Converters, used in large HVDC transmission systems are particularly prone to this. Indeed, much of the engineering design of such converters concentrates on reducing the harmonic currents to an acceptable level. It is therefore understandable that electricity utilities have expressed concern regarding the suggestion that, in future, large numbers of inverters will be used to connect PV and other renewable energy sources to distribution networks.

Fortunately, the converters used in modern renewable energy systems use techniques that reduces the low-order harmonics to negligible levels. In the context of a network supplying a typical collection of TVs and computers and so on, the harmonic contribution from renewable energy systems can be expected to be negligible.

4.5.5 The Thyristor Bridge Converter

Replacing the diodes in the bridge of Figure 4.42 with thyristors, as shown in Figure 4.26, allows the DC voltage to be controlled.

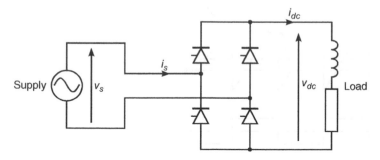

Figure 4.26 Circuit of a single-phase thyristor bridge.

In many applications, the anode-cathode current is that of a line in an AC circuit and, thus, switching-off occurs when the line current naturally reaches zero. The transfer of current from one conducting device to another previously non-conducting device is called *commutation*. Thyristors are solely suited in converter applications where the commutation process is carried out by the AC supply. Such circuits are known as *line-commutated* converters.

Notice that the load now includes an inductance, which is typical of practical applications of thyristor bridges. This inductance is usually large enough to ensure that the current i_{dc} is nearly constant, at least in the short-term from one cycle to the next. In the longer term, it will vary to match the average level of v_{dc}.

A control circuit (not shown in Figure 4.26) applies firing pulses to the thyristor gate connections. The thyristors are fired in diagonal pairs. The fired pair takeover the conduction of the current, which reduces the current in the other pair to zero and switches them off.

Control of the *firing angle* α provides a means of controlling the average DC voltage. Figure 4.27 shows the situation when $\alpha = 50°$. The shaded areas, above and below the average DC voltage are equal (because, in the steady state, the average voltage across the inductance must be zero). At $\alpha = 90°$, the average DC voltage would be reduced to zero. If α were increased past 90°, the average DC voltage would theoretically become negative. This is an impossible operating condition in the circuit of Figure 4.26, because the direct current would be required to reverse direction; that is, flow from cathode to anode through the thyristors. However, if the load resistor were to be replaced by a DC source, whose voltage was larger than the average negative voltage of the converter, the DC current direction will be maintained, power will be transferred from the DC source to the AC side and the bridge will operate as an *inverter*. Rotating the circuit diagram in Figure 4.26 through 180° puts the positive DC rail back at the top and gives the normal way of drawing an inverter, with the power flowing from left to right as in Figure 4.28.

Notice that the inductance on the DC side is still required. It causes the DC current feeding the inverter bridge to be pretty much constant (from one cycle to the next), which leads to this circuit being called a *current-fed* or *current-source inverter*.

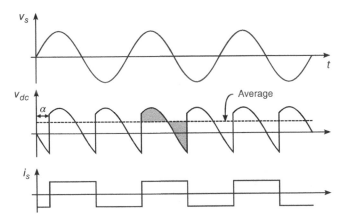

Figure 4.27 Waveforms in a thyristor bridge, while rectifying.

Figure 4.28 Line-commutated thyristor inverter.

The thyristor inverter shown in Figure 4.28 has several shortcomings:

- The current waveform on the AC side is almost a square wave (see Figure 4.27) and therefore has a very high harmonic content.
- The fundamental component of the square-wave AC current is out of phase and lagging the AC voltage. The circuit consumes reactive power from the grid.
- The circuit requires connection to the grid (or a synchronous generator) in order for the currents to commutate. Therefore, the circuit cannot be used as a general-purpose stand-alone AC supply.

Nonetheless, thyristor inverters and controlled rectifiers, based on these concepts, were the foundation of the power electronics industry during the 1960s and 1970s. Thyristor inverters were used in early variable-speed wind turbines and grid-connected PV systems.

The shortcomings of the converter based on thyristors are all but removed by the use of transistor switching devices.

4.5.6 The Transistor Bridge

During the 1980s, self-commutated power-semiconductor devices (BJTs, MOSFETs, IGBTs, GTOs and IGCTs) started to become commercially viable and this led to the development of *self-commutated inverters* an example of which is shown in Figure 4.29. In a self-commutated inverter, the switching, both on and off, of the main power-semiconductor devices is under control. Thus, a self-commutated inverter does not need a grid connection in order to operate and can be used to create a stand-alone AC power supply the frequency of which is determined by the transistor firing control circuit.

Notice that the inductance on the DC side is no longer required. This inverter is fed with a voltage that is pretty much constant (from one cycle to the next), which leads to it being called a *voltage-fed* or *voltage-source inverter*. On the other hand, for reasons explained later, an inductance is required on the AC side if the inverter is to be connected to the grid.

The diodes, shown next to each of the transistors, make the circuit appear more complicated but, in practice, this is not an issue. They are necessary in order to provide a path for the current when the transistors are turned off.

Figure 4.29 Transistor bridge.

4.5.6.1 Basic Square Wave

Applying a very simple switching pattern to the transistors in the bridge circuit shown in Figure 4.29, allows it to produce a basic square wave as in Figure 4.30.

A basic square wave can be used to supply some non-critical loads such as incandescent light bulbs and simple universal motors, but it is not suitable for the majority of AC loads, which are designed to operate from a sinusoidal supply.

The basic square wave has a sinusoidal component at the fundamental frequency, as shown in Figure 4.30, but it also has a very high harmonic content. These harmonics will typically cause excessive heating in induction motors and transformers. For the same reason, a basic square wave is not acceptable for grid connection

4.5.6.2 Quasi-Sine Wave (Modified Square Wave)

A minor modification to the control system, allows the bridge circuit to produce a quasi-sine wave.

As shown in Figure 4.31, the quasi-sine wave is really just a modified square wave, but it much closer to a sine wave. Setting the switching angles to 60–120–60–120° eliminates the third harmonic completely and makes the waveform good enough for many practical loads. The switching angles may be adjusted slightly to control the RMS voltage, bearing in mind that this adjustment increases harmonics.

4.5.6.3 Pulse-Width Modulation

For loads that are more harmonic-critical and for grid connection, *pulse-width modulation* (PWM) allows the transistor bridge circuit shown in Figure 4.29 to produce an

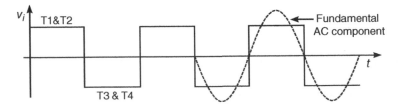

Figure 4.30 Voltage waveform of a basic square-wave inverter.

Figure 4.31 Voltage waveform of a quasi-sine-wave inverter.

Figure 4.32 PWM inverter output voltage.

almost-pure sine wave (low-order harmonics are virtually eliminated) and provides full control of its amplitude.

In a PWM inverter, the transistors are switched at a much higher frequency than that of the intended output waveform. The width (duration) of the high-frequency pulses, having short width at the edges and increasingly longer width towards the centre of the waveform as shown in Figure 4.32, is controlled so as to create a good approximation of a sine wave output.

In order to provide a high-quality sine wave, the switching frequency needs to be as high as possible. But if it is too high, however, *switching losses* will become significant, making the inverter itself less efficient. Transistors, when at their on or off state, dissipate very little power. It is during the periods of transition between the two states that the devices are most lossy. For small MOSFET PWM inverters, switching frequencies up to 20 kHz are typical. While PWM virtually eliminates the low-order harmonics, there can be significant harmonics around the switching frequency and its multiples, but these can readily be filtered out.

Figure 4.33 illustrates one simple method that the inverter's internal control system may use to create the switching signals. The pulses that switch the transistors on and off are generated at the intersections of the reference sine wave with the carrier wave that is usually triangular. In a stand-alone inverter, the reference sine wave would be created by the internal control system, normally to provide a constant voltage and constant frequency output. Reference [3] provides more information on this rather complex topic.

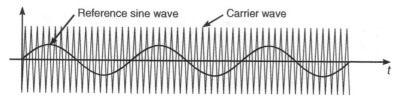

Figure 4.33 PWM construction of switching pattern.

4.5.6.4 Comparison of Switching Methods

Quasi-sine wave inverters are widely used for small stand-alone applications. For larger applications, and for grid-connected inverters, PWM is normally employed.

This discussion has focused on inverters providing AC power at 50 or 60 Hz, but inverters are also used to provide power at much higher frequencies. This is very useful internally in converter systems because it allows the use of high-frequency transformers, which, as mentioned earlier, are much smaller, lighter and cheaper than 50/60-Hz transformers of the same power rating. This is a topic that will be visited later when converters for renewable energy sources will be reviewed.

4.5.6.5 Output Control in a Grid-Connected Inverter

In a grid-connected inverter, the reference sine wave shown in Figure 4.33 is created so that it is of the same frequency as that of the grid but it can be phase-shifted with respect to the grid voltage.

Referring to Figure 4.29 and using phasors rather than instantaneous values, it is clear that:

$$V_i = V_o + V_x$$

where V_i is the fundamental component of the pulse-width modulated voltage output from the inverter bridge, V_o is the output voltage, which is the grid voltage once the inverter is grid connected, and V_x is the voltage across the inductor.

Furthermore, the voltage across the inductor must lead the output current I_o by 90°, that is:

$$V_x = jI_oX$$

where X is the inductor reactance.

Phasor diagram (a) in Figure 4.34 illustrates the general case and shows that control of the amplitude and phase of V_i with respect to the grid voltage V_o provides control of both amplitude and phase of the output current I_o. This is essentially the mechanism for the control of the active and reactive power flows as described in Eqs. (4.8c) and (4.8d) in the case of the synchronous generator.

Phasor diagram (b) shows the particular case where the inverter is operating at unity power factor (zero reactive power flow) where the output current I_o is in-phase with the grid voltage V_o. The required amplitude of V_i and its phase with respect to the grid voltage can readily be calculated for any required output current I_o.

Furthermore, neglecting losses in the circuit, it is known that power-out must equal power-in, that is:

$$V_{dc}I_{dc} = V_oI_o \cos \phi$$

 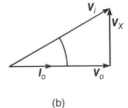

Figure 4.34 Phasor diagrams of output stage of a grid-connected inverter; the (a) general case and (b) at unity power factor.

(a)   (b)

Figure 4.35 Three-phase IGBT bridge.

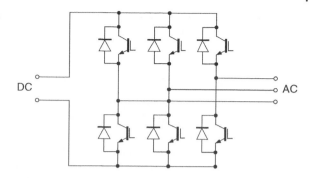

where I_{dc} is the average value of the current drawn from the DC source. Thus, by using the PWM pattern to control the output, the input current can be controlled.

4.5.6.6   The Three-Phase Bridge

All of the preceding circuits and discussions may readily be extended to three-phase systems. Usually, this only requires an additional 50% of devices as shown in Figure 4.35 where a single-phase bridge with four transistors has become a three-phase bridge with six transistors. Inverters above 10 kVA are usually three-phase.

Three-phase power electronic converters are extensively used in variable-speed wind turbines to convert AC to DC and vice versa. The circuit shown in Figure 4.35 is known variously as a transistor bridge, an inverter, a voltage-source converter, a variable-frequency drive or just about any combination of those words. It is widely used throughout industry to operate induction motors at variable speed. IGBT converters are expensive, typically several times the cost of the associated electrical machine. These are known as two-level converters because the power electronic devices switch between two distinct voltage levels. More recently, multi-level inverters are becoming common and these are discussed in Section 4.5.9.

4.5.7   Converter Internal Control Systems

Very simple converters may employ analogue electronics to provide the transistor gate or base control signals. In all other cases, the internal control of a modern converter is provided by a dedicated microprocessor or digital-signal-processor (DSP). By controlling the exact timing of transistor gate or base control signals, the microprocessor has control over the currents, voltages and power flow through the converter. The microprocessor will often be used to provide a number of other functions. In a grid-connected PV inverter, for example, the microprocessor will typically perform the maximum power point tracking (MPPT), anti-islanding protection and perhaps some diagnostics and communications.

Depending on the application, the microprocessor-based control system in a modern converter may have a variety of control objectives. For example, in a stand-alone inverter, the objectives would be to keep the output at constant voltage and constant frequency; in a grid-connected PV inverter, the objective would be to achieve maximum power extraction as the solar radiation varies, a task that would require control of the inverter output current and in a motor drive inverter, the objective might be to control the torque.

The power electronic circuits used in all three of these examples could be very similar, but the implemented control systems make them behave very differently.

4.5.8 DC–DC Converters

An AC voltage can be easily stepped up or down through the use of the ubiquitous transformer. Unfortunately, this facility was not as easily available for DC until the recent explosion in power electronic developments. As will be shown later, a DC transformer widens considerably the application of power electronic converters.

An ideal DC–DC transformer will convert power at one voltage to power at a different voltage. If the voltage is increased, the current will be decreased by the same factor, and vice versa, such that:

$$V_{in}I_{in} = V_{out}I_{out} \tag{4.29}$$

DC transformers are known as *choppers*, or *DC–DC converters*. In such choppers, a power transistor is switched at high-frequency and the *duty cycle* (the on-to-off time ratio known as the *mark-space ratio*) controls the output voltage.

4.5.8.1 Step-Down DC–DC Converter

Provided that the current in the inductor is continuous, the steady-state operation of the circuit in Figure 4.36 can be described as follows:

The transistor is switched on for time T_{on}, during which $v_l = v_{in} - v_{out}$. Then it is switched off for a time T_{off}, during which $v_l = -v_{out}$. In the steady state, the volt-seconds across the inductor must be zero, so:

$$V_{out} = DV_{in} \tag{4.30}$$

where D is the duty cycle: $\dfrac{T_{on}}{T_{on} + T_{off}}$

4.5.8.2 Step-Up DC–DC Converter

Provided that the current in the inductor is continuous the steady-state operation of the circuit in Figure 4.37 is given by:

$$V_{out} = \frac{1}{1 - D}V_{in} \tag{4.31}$$

Figure 4.36 Step-down DC-DC converter (also known as a series switching regulator or buck converter).

Figure 4.37 Step-up DC-DC converter (also known as a boost converter).

4.5.9 Multi-Level Converters

Multi-level converters have been under development over the last decade and are now finding widespread application, including for wind power due to the increase in the rated power of wind turbines. Their main advantage lies in improved power quality, and their capability to deliver higher power ratings. The total harmonic distortion (THD) decreases as the number of voltage levels increases. Different configurations of multi-level converters are possible; the most common being the neutral-point-clamped converter and the cascaded H-bridge inverter (basically a group of two-level H-bridge converters connected in series), as shown in Figure 4.38. PWM strategies have been extended to control these multi-level converters. For wind turbines in the range 3–7 MW with full converters, three-level converters would appear to be the most cost effective.

4.5.10 Matrix Converters

Matrix converters are connected directly between two different AC systems where any input phase can be connected to any output phase; to do this, they make use of an array of bidirectional power electronic switches. There is no DC link, and thus no need for a large storage capacitor. The key component is a fully controllable four-quadrant bidirectional device that is capable of high-frequency switching. For three-phase systems, a minimum of nine bidirectional switches is required. Such an arrangement is shown in Figure 4.39. These bidirectional switches are formed from a combination of conventional power electronic switches and diodes. Specific control algorithms have been developed for these converters since the 1980s.

To date, commercial matrix converters have been limited to around 1 MW, and have generally been applied in smaller units. They have yet to find any significant application in renewable energy systems.

4.5.11 Z-Source Converters

A Z-source converter is an inverter that can either buck or boost the input voltage by the use of passive components. This is done using an LC impedance circuit to couple

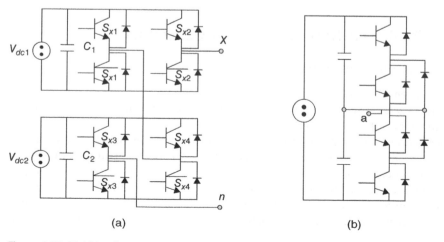

Figure 4.38 Multi-level converter topologies: (a) cascade connection and (b) diode-clamped.

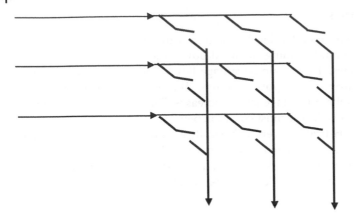

Figure 4.39 A generic three-phase matrix converter topology.

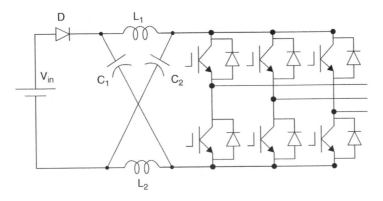

Figure 4.40 Basic configuration of a Z-source inverter.

the converter to the power source. For applications that require a DC/DC boost stage, such as PV systems (see next), the boost converter can be eliminated and this improves the overall conversion efficiency and reduces the cost (Figure 4.40).

4.6 Applications to Renewable Energy Generators

4.6.1 Applications to PV Systems

4.6.1.1 PV System Characteristics

All PV systems generate DC and so power electronic converters, known as inverters, are required to connect them to power systems.

As described in Chapter 2, PV modules have an IV characteristic that depends on incident radiation intensity (and to a lesser extent the solar spectrum) and module temperature. Different types of PV have different shapes of IV curve, characterised in simple terms by a fill factor, as explained in Chapter 2. To maximise the output of any PV module, or collection of modules, operation should be at the maximum power point. This point will change with radiation and temperature and a key function of the inverter is to

Figure 4.41 Basic grid-connected PV inverter.

track this. How effective this so-called MPPT is can critically effect the overall efficiency of the PV system.

4.6.1.2 Basic Grid-Connected PV Inverter

The grid-connected PV inverter shown in Figure 4.41 is based on the transistor bridge circuit of Figure 4.29. It is required that the DC voltage of a PV array is inverted and connected to a grid of say, 230 V. For such a simple, domestic application, the array voltage would have to be well over 350 V DC at the maximum power point, which would be impractical in many cases. The transformer is therefore required in order to raise the output inverter voltage.

The control could be implemented as described earlier under the heading *Output Control in a Grid-connected Inverter* in Section 4.5.6. Control of the AC output current would provide control of the DC current drawn from the array and hence allow MPPT. The main drawback of the circuit is the size, weight and cost of the output 50 or 60 Hz transformer.

4.6.1.3 Transformerless Grid-Connected PV Inverter

The addition of a DC booster stage, at the input of the inverter, as shown in Figure 4.42, removes the need for the mains frequency transformer, and reduces the current in the inverter transistors as they operate at a higher voltage than in Figure 4.41.

The converter in Figure 4.42 has two stages, both of which need to be controlled.

One possibility would be to operate the DC booster at a fixed duty cycle, which would give a fixed ratio between the average DC currents and allow the MPPT to be implemented by controlling the AC output current as it was in the previous example.

In practice, it may be more efficient to fix the voltage of the DC bus (the DC connection from the booster to the inverter). In this case, the control of the inverter current is used

Figure 4.42 Transformerless grid-connected PV inverter.

to provide control of the DC bus voltage – if the voltage is too high, more current is drawn, and vice versa. MPPT is then implemented by controlling the duty cycle of the DC booster.

A perceived drawback of this converter is that it provides no electrical isolation between the grid and the PV array. This means that the array is at mains voltage and that for safety requirements must be double insulated. Some commercial inverters currently being manufactured are based on this arrangement.

4.6.1.4 PV Inverter Using a High-Frequency Transformer
The inverter shown in Figure 4.43 has three stages. Starting from the left, the first is a high-frequency inverter, which could be a quasi-square-wave inverter, or a square-wave inverter. Operation at 20 kHz is typical and this allows the transformer that follows to be much smaller, lighter and cheaper than the mains transformer used in Figure 4.41. The transformer steps up the voltage and provides the isolation that was missing in the circuit Figure 4.42. Next, the high-frequency AC is rectified to provide DC suitable for the final stage PWM inverter on the right of the diagram.

Although this circuit appears complicated, the cost reduction of the transformer makes it attractive to manufacturers.

4.6.1.5 PV Inverter Using a Steering Bridge
The inverter shown in Figure 4.44 again makes use of a high-frequency transformer, but differs from that of Figure 4.43 in that the power flow through the transformer is now modulated at 50 or 60 Hz. This modulation is achieved by use of a more sophisticated switching pattern on the input (left-hand) inverter and is designed to make the current i_x take the shape of a rectified 50 or 60 Hz sine wave. The output (right-hand) bridge

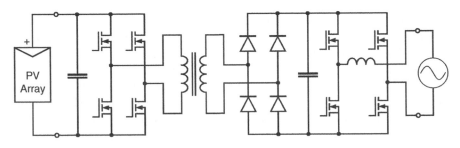

Figure 4.43 PV inverter using a high-frequency transformer.

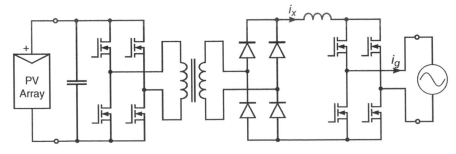

Figure 4.44 PV inverter using a steering bridge.

then 'un-rectifies' or 'steers' this to produce a full sine wave current in to the grid. The advantage of this, over the arrangement shown previously in Figure 4.43, is that the steering bridge switches at just 100 or 120 Hz and only when the instantaneous current is zero, making it more efficient than the PWM inverter on the right of Figure 4.43.

4.6.1.6 PV Inverters for Stand-Alone Operation

PV is often used to supply loads in remote areas, well away from grid. Indeed, perhaps the most common application of PV is to supply essential domestic loads; these are known as solar home systems. They commonly involve battery storage. The PV module/s are connected to the battery, sometimes directly, and sometimes via a DC–DC converter, as described in Section 4.5.8, to track the maximum power point of the PV supply. AC loads are then supplied from the battery using a small voltage-source inverter, which can be quite basic in design (e.g. the simple square-wave inverter described in Section 4.5.6).

4.6.2 Applications to Wind Power

4.6.2.1 Fixed Versus Variable Speed – Energy Capture [4]

This section deals with average changes of wind speed over tens of seconds or minutes and the wind turbine response to maximise energy capture over this time scale. In Section 2.4.3, mention was made that for maximum energy extraction the wind turbine speed should be adjusted to vary in sympathy with the wind speed. Figure 4.45 shows typical wind turbine power versus rotational speed characteristics with wind speed U as a parameter with $U_1 < U_2 < U_{rated}$. For each curve the converted power is

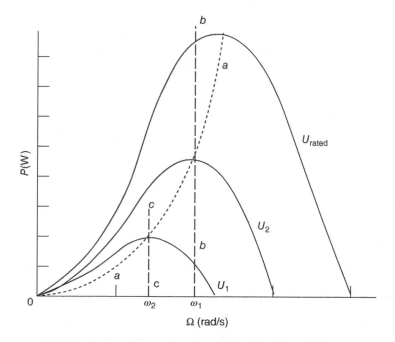

Figure 4.45 Wind turbine power-rotational speed relationship with wind speed as a parameter.

zero at standstill ($\omega = 0$), and at runaway speed when the turbine is rotating without any restraining torque. In between these two extremes, the power rises to a maximum. It stands to reason that for each wind speed it is desirable to operate the wind turbine at this maximum power point. For this to be possible as the average wind changes, the turbine torque must be adjusted so that the locus aa that passes through the maximum power points is tracked. This is an alternative way of expressing the requirement stated in Section 2.4.3 where the tip speed ratio $\lambda = \Omega R / U$ should be fixed at λ_{opt}, the value at which C_p is a maximum. For each wind turbine design, λ_{opt} is different; for example, in Figure 2.9, C_p is a maximum when $\lambda \approx 7$.

Unfortunately, synchronous generators and to all intents and purposes cage rotor asynchronous generators, operate at fixed speed, say at ω_1 in Figure 4.45, and a turbine driving either generator would track a locus such as 'bb'. Only at wind speed $U2$ would the wind turbine operate at λ_{opt} with the consequence that for higher and particularly lower wind speeds less than maximum energy would be extracted. Operation at variable speed (to track locus bb) with either synchronous or asynchronous generators is possible through the use of power electronic converters. Due to the losses of such converters the additional energy captured by a wind turbine operating at variable rather than fixed speed is typically less than a few percent and would not alone justify the extra expense. However, there are other reasons, mainly concerned with loads, why variable-speed operation is desirable.

4.6.2.2 Fixed Versus Variable Speed – Dynamics

The vast majority of the world's electricity is generated by synchronous machines directly connected to their respective power systems. This configuration works very well when the prime mover (usually an engine or a steam, gas or water turbine) provides a steady torque to the generator. Wind, however, is turbulent and this translates directly into fluctuations in drive train torque. Using a directly connected synchronous generator in a wind turbine would form too rigid a coupling between the mechanical and electrical systems. Wind gusts would cause large mechanical stresses in the turbine and, depending on the nature of the electrical grid, large fluctuations in the power fed into the electrical system.

The main advantages of variable-speed wind turbines in terms of dynamics are:

- The total inertia of the aerodynamic rotor, the gearbox (if there is one) and the electrical generator act as an energy buffer, smoothing out the wind turbulence. Transient torques and rapid variation in electrical power as well as stresses in the drive train are greatly reduced
- Lower structural loads and lighter foundations, are other advantages of particular importance in offshore applications

Additional advantages are:

- The power electronics may also be capable of regulating the reactive power flow in the network
- Noise is reduced when operating at low wind speeds.

The drawbacks of variable speed are the extra complexity of the generator (in some schemes) and of the necessary power electronic hardware all of which increase cost and possibly reduce reliability.

To allow for wind gusts, virtually all wind turbines have generators or generator systems that provide some degree of speed variation.

4.6.3 Synchronous Generator Supplying an Autonomous Network

A synchronous generator driven by a turbine through a gearbox is exclusively used to supply an autonomous network. In such networks, the wind turbine may be the only source of power or more often may be supported by a storage system; for example, a battery plus converter or another source such as a diesel driven generator. At present, the relatively few examples of this scheme are restricted to small island or to remote communities. If the wind turbine is the main source of energy, it is also in charge of system frequency and voltage. For a fixed frequency grid, the wind turbine has to be controlled to operate at fixed speed in spite of variations in wind speed and in system demand. This can be achieved through (i) a stall regulated wind turbine with frequency sensitive prioritised load switching or (ii) a combination of variable pitch wind turbine control plus a controllable dump load and/or a storage system such as batteries with a bidirectional converter. Such autonomous systems are at present unusual and will not be referred to further.

4.6.3.1 Fixed-Speed Wind Turbines

Fixed-speed wind turbines utilising a gearbox and a direct grid-connected asynchronous (induction) generator were the first to be widely used and were dominant through the 1980s. Figure 4.46 shows this classical arrangement with the robust and low cost squirrel-cage generator. It is simple, reliable and very well proven in practice. Many of the wind turbines installed during the 1980 and 1990s were of this type and can still be found in service today. These turbines were initially stall regulated in that the fixed-speed rotor would progressively stall, regulating power, as the wind speed increased, but later direct connected would use pitch regulation rather than stall. In this scheme the induction machine is of cage configuration and due to the inevitable slip, the speed varies marginally with power so the speed is not strictly constant. The small mechanical *compliance* due to the slip provides some degree of electrical power smoothing during gusty conditions and is of benefit to reducing loads in the mechanical drivetrain. Such fixed-speed wind turbines are now known as Type I turbines according to the IEC classification.

Figure 4.46 Directly connected squirrel-cage induction generator.

Where a large induction machine is to be directly connected to the network (without speed control), some form of *soft starter* is generally required in order to limit the currents and torque during start-up. A soft starter is not a converter as such but simply limits the surge of current necessary that an asynchronous generator absorbs for the establishment of the RMF when first connected to the electricity network. As it uses thyristors it is much cheaper than the transistor-based interfaces.

Last, Figure 4.46 shows a transformer that may be located in the nacelle or in the base of the tower and may or may not be included in the turbine manufacturer's scope of supply. The induction generators in wind turbines usually operate at 690 V and the transformer steps this up to the voltage of the windfarm power-collection network (typically 11 kV or 33 kV in the UK). Large windfarms may have a further transformer to step-up the voltage again before connection to the utility grid.

4.6.3.2 Variable Slip Wind Turbines

The squirrel-cage rotor, used in the fixed-speed wind turbines described previously, is very economic, robust and reliable. It has a small but finite resistance, which largely determines the slope of the torque-speed characteristic and hence the magnitude of the slip. Larger squirrel-cage induction machines generally have proportionally lower rotor resistance and therefore less slip. This reduces the compliance between the mechanical and electrical systems, which is not desirable in wind turbines. Figure 4.16 illustrates one way in which the speed of the induction machine could be regulated. The rated torque can be developed at different speeds by increasing the rotor resistance from Rr through to Rr' and then to Rr''. This can be easily achieved by introducing an external set of variable resistors across the slip rings of a rotor wound induction machine. Controlling the speed in this way, however, incurs an unacceptable reduction in efficiency due to the energy dissipated in the external resistors. This mode of control could be dismissed as useless in wind turbines; nevertheless, a manufacturer has used it intelligently and successfully to improve performance.

In a wind turbine, the resistance can be controlled dynamically to allow for wind gusts. During the gust, the resistance is increased, allowing a little more (negative) slip; the rotational speed increases slightly and energy is absorbed by the inertia. After the gust, the resistance is reduced, which brings the speed back down and exports the energy to the grid over a longer period of time. Figure 4.47 shows such a scheme. If this scheme were to be applied to wind speeds between cut-in and rated, valuable

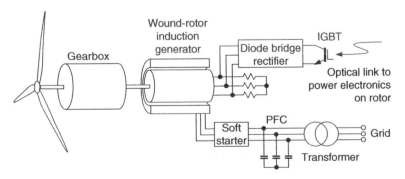

Figure 4.47 Vestas OptiSlip rotor resistance controller.

energy from the wind would have been irreversibly dissipated in the external resistors. However, if this technique is applied at times when the average wind speed is above rated, that is when energy in the wind can be shed without economic penalties, an interesting benefit results. In what follows, the rationale of this policy will be explained.

In the Vestas OptiSlip scheme [5, 6], a specially designed asynchronous generator with a wound rotor is fitted with rotating integrated electronics and resistors on the shaft thus eliminating the need for slip rings and brushes. Here, the blade pitch control mechanism takes care of the longer variations in the wind, while the OptiSlip system looks after the faster variations due to gusting. The power output is maintained at the rated value with the speed allowed to vary between the 1 and 10% slip range. The variable slip concept produces high-quality electrical output at high wind speeds without generating any detectable harmonics. The variable slip concept does not enhance energy capture but improves the quality of generated electricity and the stresses in the wind turbine drive train in the region that matters, that is, for wind speeds above rated. Such turbines, with limited speed variability, are known as Type II turbines.

4.6.4 The Principle of Slip Energy Recovery

A far superior approach to dissipating energy in external rotor resistors is to capture this energy and transfer it back into the grid. There are a variety of such *slip energy recovery* schemes. In this section the general principles of slip energy recovery will be outlined before dealing with its application to wind turbines. At the outset it must be remembered that the frequency and magnitude of the induced voltages and the resulting currents in the rotor circuit of an asynchronous generator depends on the slip. Transferring energy out of or into the rotor circuit requires the interfacing of this variable-frequency variable magnitude rotor voltage to the fixed frequency, fixed magnitude voltage of the mains. This can be achieved through an *asynchronous link* consisting of an AC to DC converter followed by a DC to AC converter.

The technical problems in such a scheme are not trivial. At very low slips the rotor side converter will have to deal with very low voltage, low frequency rotor voltages. Conversely, at high slips, the voltage and frequency will be proportionately larger. It can be shown that for control of speed from zero to synchronous, the rating of the rotor-connected DC link will have to be the same as the nameplate rating of the induction generator. This would involve an electronic interface of considerable size and therefore cost. The advantages offered by full variable-speed control can be almost matched if a limited range of control is provided. This can be ingeniously arranged by a rotor-connected asynchronous link of relatively low rating. To investigate this arrangement, the possible power flows in and out of the rotor of an asynchronous machine need to looked at more closely.

Figure 4.48a shows the power flow in an induction machine when motoring. The electromagnetic power P_{ag} crossing the air gap is equal to the electrical power P_s fed into the stator minus the losses referred to in Section 4.4. In turn, subtracting the rotor losses from P_{ag} gives the useful mechanical power output P_m. Neglecting the losses, the simplified diagram of Figure 4.48b can be set up with the aid of Eq. (4.22). The asynchronous machine behaviour when the machine is generating will now be examined, and with the assumption that the capability exists of extracting and injecting power into the rotor circuit through an asynchronous link. In other words, there is an ability to adjust the

(a)

(b)

Figure 4.48 Power flows in and induction machine. (a) Power flows when motoring and (b) simplified power flows.

magnitude and sign, that is, the direction of P_r in Figure 4.48b. In the figure, the arrows indicating power flows are shown in the conventional way from left to right, that is, when the machine is motoring.

In the usual operation as a generator (Figure 4.16) the slip is negative in the range $-1 < s < 0$, and P_m flows from right to left, that is, has a negative sign. As a consequence, P_{ag} and P_s have also reversed sign but P_r is positive. By extracting the required level of P_r from the rotor circuit, the induction generator speed can be controlled in the *supersynchronous* range. In this mode, the mechanical input power is fed into the power network partly through the stator and partly through the rotor.

Through the asynchronous link, the capability exists of injecting rather than extracting power into the rotor circuit, thus reversing the sign of P_r. The consequence of this is that as P_m and P_{ag} are negative the slip now must be positive and in the range of $1 > s > 0$. The induction generator can therefore be made to operate at a *subsynchronous* speed. In this mode of operation, $P_m < P_{ag}$; hence, the electrical power fed into the rotor from the electricity network is returned to the network through the stator. Due to the possible bidirectional flow in the rotor circuit this machine is often referred to as being *doubly fed*.

It can be shown that the rating of the asynchronous link in the rotor circuit is proportional to the required speed shift from synchronous, for example, a speed shift of 30% will require an asynchronous link rated approximately at 30% of the nameplate rating of the induction generator. The advantage of modifying slip energy should now be obvious. By having a bidirectional link capable of extracting or injecting 30% of the rated power in the rotor circuit, a 60% speed variation can be achieved. For the purposes of reaping the advantages of variable-speed operation this is almost as good as full speed control. It is not surprising that a number of major manufacturers of wind turbines have chosen this technique for their multi-megawatt wind turbines.

4.6.4.1 DFIG Wind Turbines

DFIG stands for a *doubly fed induction generator* [4]. Figure 4.49 shows a simplified schematic of a wind turbine connected to a doubly fed induction generator. The rotor-connected voltage-source converter IGBT converter is linked to a grid-connected IGBT converter through a DC link containing a capacitor. Both converters are capable of *four-quadrant* operation; that is, they can be controlled to transfer active and reactive power in either direction. The rotor-connected converter has a more onerous task to perform than the grid-connected one.

The rotor-connected converter looks after the wind turbine speed control so that energy extraction is maximised. The generator is controlled on the basis of *vector control* a complex technique that requires substantial on-line computational requirements. This technique 'decouples' the active and reactive power control functions permitting the induction generator to be controlled with exceptional flexibility. It also offers superior transient response permitting the induction generator to follow very rapidly commands on torque change. This enhances the energy extraction capability from gusts and relieves drive train stresses. For this computation to be executed, it is necessary to feed into the converter controller information on the rotor speed and its angular position, the magnitude and phase of the rotor and stator currents and information on the torque-speed curve that must be followed to maximise energy extraction.

Reactive power, or power factor, could also be externally specified. Although either converter can be arranged to inject reactive power into the network (the former through the rotor-stator path, the latter directly), it was found that the task can be more economically performed through the rotor-connected converter.

The function of the grid-connected converter is, in most schemes, to transfer the required active power from the DC link to the network or vice versa. This is carried out through a voltage control loop, which ensures that the DC voltage is maintained fixed. For example, if the rotor-connected converter decides that more energy should be transferred from the WT to the DC link (as a response to an increase in wind speed), the capacitor will be overcharged and its voltage will increase. This will prompt the grid-connected converter to transfer more energy from the DC link to the network thus restoring voltage equilibrium. Because of the substantial variation in the magnitude of the voltage in the rotor of the asynchronous generator, the DC rail voltage is kept low. For an economic design this necessitates the connection of the converter to the grid through a transformer.

Both converters are controlled through PWM techniques that ensure minimum harmonic injection into the electricity network. To protect the power electronics, an overvoltage protection is mounted on the rotor terminals to short circuit the rotor in case of equipment malfunction. The converters in Figure 4.49 are rated to carry only the rotor

Figure 4.49 DFIG (doubly fed induction generator).

Figure 4.50 Wind turbine with induction generator and full converters.

power, typically 30% of the wind turbine rating. Wind turbines in which a limited proportion of the power pass through power electronics, such as DFIGs, are known as Type III turbines.

Technical details of the various wind turbine generator concepts presented here can be found in Reference [7].

4.6.4.2 Wind Turbines with Full Converters

The configuration shown in Figure 4.50 uses two full converters. Full, in this context, means that all the power goes through the converters. Such converters allow the standard squirrel-cage induction machine to operate at variable frequency and therefore 0 to 100%. variable speed. Full variable turbines can also use synchronous generators; the full refers to the converter. All such turbines are known as Type IV.

The active power from the induction generator is rectified by the machine-side converter and fed to the DC link. The grid-side converter acts as an inverter, taking power from the DC link and feeding it to the grid.

The structure of this link is very similar to that of Figure 4.49. Vector control of the induction generator is again implemented. The reactive power demand of the induction generator is supplied by the rotor-connected converter, while the grid-connected converter has the task not only of transferring the whole of the active power generated to the power network but also of injecting or absorbing reactive power to/from the power system, respectively. Few manufacturers have adopted this option, which requires a cheaper machine but a more expensive DC link compared to the DFIG.

The converters used in this scheme are standard industrial drives, as would normally be used for variable-speed motors; they are reliable and readily available from numerous manufacturers. The main drawback of the configuration is that it requires two IGBT converters rated at the full power of the wind turbine and these are expensive. The benefit is that the generator is a very robust squirrel-cage machine.

4.6.5 Synchronous Generators in Wind Turbines

As noted earlier, the main reason for using an induction generator in a wind turbine is that for fixed speed, the slip provides some compliance and non-oscillatory response (Table 4.2), and for variable speed, the ingenious rotor power recovery scheme provides economic speed control. However, in the case of a fully variable-speed wind turbine a synchronous generator may again be considered.

In this context, a synchronous generator has two advantages:

- Unlike an induction generator it does not need a supply of reactive power and therefore the output can be connected to a DC bus through a simple bridge rectifier using diodes or thyristors.
- It can be manufactured with a very large diameter and large number of poles. Such a machine has a low operating speed and high torque, and can be designed for direct mechanical coupling to the rotor of a wind turbine, eliminating the need of the gearbox.

A synchronous generator can be constructed either with a wound rotor carrying a controllable DC field excitation current or with a permanent magnet rotor. Both approaches have been demonstrated in large-scale wind turbines. A wound rotor has the advantage of controllability of the generated voltage through field current control. This reduces the DC link duties compared to rotor with permanent magnets in which the generator voltage is just proportional to the speed. However, with permanent magnets the generator is more efficient due to the absence of the excitation loss. In either case, the variable-frequency AC in the machine stator need not be sinusoidal and the freedom to use a trapezoidal waveform, for example, can be exploited in electromagnetic design of the machine. Additionally, as the AC output is rectified, it is not required to generate a frequency as high as 50 Hz at the highest rotational speed.

4.6.6 Gearless Wind Turbines

Use of a synchronous generator through a full converter also reopens the possibility of eliminating the gearbox. This approach has been successfully followed by the company Enercon since 1993. The first consideration in designing a gearless wind turbine is that the rated torque of any electrical machine is roughly proportional to the volume of its rotor. The gearbox in a typical 1 MW wind turbine has a ratio of about 70. Eliminating the gearbox increases the torque and hence the volume of the required generator rotor by roughly the same factor. To achieve the volume without the weight, and to achieve a sufficient rate of flux cutting, a very large-diameter ring-shaped generator must be used. It is still considerably heavier than the usual 1500 rpm generator found in most wind turbines, but Enercon consider that this is justified by the elimination of the gearbox.

The Enercon ring generators have a large number of poles: 72 in the case of the 1.5 MW E-66. The rotational speed of between 10 and 22 rpm thus produces a frequency of between 6 and 13.2 Hz. Also, their generators are six-phase, which helps to reduce the ripple on the DC coming out of the diode bridge rectifier. The gearless configuration has been used in all Enercon models from the 300 kW E-30 in 1993 to the new 4.5 MW E-112.

As shown in Figure 4.51, the machine-side converter could be a simple diode rectifier and this is typical in small wind turbines. In larger wind turbines, a controlled rectifier is useful. Depending on the choices before, the DC bus may or may not keep a constant voltage as the speed varies. The grid-side converter is likely to be an IGBT based inverter that is required to transfer active power unidirectionally but is capable of injecting or absorbing reactive power from the grid.

More recently, as part of the significant scale up of turbines for offshore application, a number of direct drive wind turbines using permanent magnet generators have been

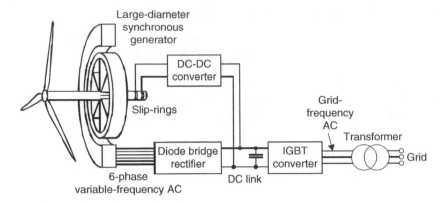

Figure 4.51 Gearless wind turbine using synchronous generator.

developed. In particular, Siemens developed a 6 MW turbine, the SWT-6.0-154 with 154 m diameter rotor, based on an in-house permanent magnet generator design that is significantly lighter that the equivalent conventional drive train with gearbox. Two 3-MW converters (developed for their earlier 3-MW direct drive wind turbine) are used in parallel. Three turbines were installed at the Scottish and Southern Energy (SSE) test site at Hunterston on the west coast of Scotland in 2013. Since then Siemens have uprated this design to produce the 7-MW SWT-8.0-154 and the 8-MW SWT-8.0-154 turbines, using the same rotor, all aimed at the offshore market.

Alstom have also developed a 6MW direct drive turbine with a rotor diameter of just over 150 m. This has been under test offshore at the Belwind 2 site off the coast of Belgium since November 2013. Series manufacture is expected to start in 2015 to supply turbines to EDF's first offshore wind farms to be constructed of the French coast (EDF's first offshore wind farm is located just offshore near Hartlepool on the north-east coast of England but uses smaller 2.3 MW Siemens turbines).

These large full converter wind turbines are likely to use multi-level converters to meet the power requirements. An excellent review of the different configurations of such converters, and their application to wind turbines can be found in Reference [8].

4.6.7 Hybrid Drive Train Designs

Not all manufacturers are convinced that the direct drive solution is the most cost effective and a number have developed so called hybrid drive trains that combine gearboxes with a reduced number of stages compared to conventional geared drive trains with medium-speed multi-pole permanent magnet generators and full converters.

The Spanish manufacturer Gamesa was one of the first to manufacture large turbines with a two-stage gearbox. Their 5MW design (model G128-5.0) combines a 128 m rotor with a two-stage gearbox and a permanent magnet generator; serial production began in 2014.

Samsung Heavy Industries have produced a 7 MW hybrid drive train turbine: the 171 m diameter S 7.0-171. This was installed at the Fife Energy Park in Scotland in 2013. It uses a medium step-up gearbox and a 400RPM permanent magnet generator.

Areva announced a medium-speed 8 MW design for offshore application in November 2013. Following input from Gamesa, this has now become the Adwen AD 8-180 and features a rotor diameter of 180 m, a medium speed, two-stage step-up gearbox and a permanent magnet generator. The first prototype was installed in Bremerhaven and commercial production is expected from 2018.

Only time will tell whether direct drive schemes or 'hybrids' will eventually supplant the well-established technology of variable slip induction generators.

Technical details of the various wind turbine generator concepts presented in this section can be found in References [9–11]. Reference [12] provides an excellent review of turbine design principals including drive trains.

4.6.8 DC Transmission for Wind

AC cables possess large capacitance resulting in large reactive power currents that are proportional to cable length. As a result, the useful power that can be transmitted reduces as length and voltage increase, with viable AC cable lengths limited to only a few tens of kilometres. Cables when run with DC have no length limit. Additionally, they are cheaper and have reduced losses. DC schemes also provide a better level of reliability if a single cable is lost.

In the longer term, as wind farm sizes grow, perhaps to several GW, and the distance to shore increases, it is likely that DC transmission may be used to link the wind farms to the grid network. The cost of a DC line to transmit a certain power is considerably lower than that of the AC equivalent. However, generators are most efficient when designed to provide AC, and AC is unbeatable when used to distribute energy to consumers at different voltage levels. Hence, DC transmission requires conversion from AC to DC and from DC to AC at the extremities of the DC line. This is achieved through power electronic converters.

DC transmission is used extensively at present in power networks when bulk power has to be transmitted over long-distance overhead lines or medium distance underground or undersea cables. It can be shown that if the power to be transmitted is large enough and the overhead or cable line is above a critical length, the savings in transmission line costs more than compensates for the extra expense of the converters. Because of the DC based coupling of the AC networks, a DC link is asynchronous; that is, the two linked AC systems do not have to operate at the same frequency. This was illustrated in Figure 4.51 where a gearless wind generator is connected to the network through a DC link. In conventional HVDC (high voltage DC) links the converters used are the line-commutated as shown in Figure 4.26. This technology has a proven track record but also suffers from the disadvantages listed in Section 4.5.5.

As indicated previously, HVDC is not just used for offshore wind. In countries like China, where energy sources (including wind) are often located far from the major load centres, HVDC is in common use. So much so that there are concerns about the reliability of the line-commutated inverters in the event of voltage disturbances near to the converters than can upset their commutation, causing them to trip off. Nearby to Shanghai a number of HVDC lines terminate and there is the possibility of cascade tripping. China State Grid, who operate these systems, is now considering the installation of synchronous condensers (large but unloaded spinning synchronous generators permanently connected to the network) in order to help stabilise the voltage waveform.

An HVDC link can be realised using the voltage-source converters (VSC) described in Section 4.5.6, rather than line-commutated converters. This technology has only become possible through recent advances in high-power IGBTs. Unlike line-commutated converters, VSCs have no need for an AC source commutation voltage, are capable of independently controlling the active and reactive power and inject negligible amounts of harmonics into the AC network. Another advantage of VSC-based HVDC is that it can be built on a modular basis using standardised units to build up a converter station. It is because of these characteristics that utilities are usually in favour of VSC-based HVDC technology. Compared to the AC option, the VSC-HVDC would be substantially more expensive in near to shore wind farm schemes but would become more favourable with increasing scheme sizes and distances.

4.7 Applications to Small Scale Hydro

Small scale hydro schemes can be grid connected, in which case the variable-speed drive arrangement can be very similar to those used for wind turbines. The motivation though is rather different as water flow rates vary very slowly, and there is no equivalence of short-term turbulent wind variation and associated loads that need to be mitigated. Large-scale hydro-turbines do not require variable-speed drives as they are operated at fixed speed with adjustable turbine blade pitching and controlled guide vanes used to regulate output. However, such technology is not cost effective for small scale turbines that can exhibit a significant variation in the optimal rotor speed with water flow rate. Thus, the motivation for variable-speed operation here is to optimise operation in response to slowly changing water flow rate. Figure 4.52, taken from Reference [14], shows how measured power from a small hydraulic turbine (a modified Kaplan turbine with no pitch or guide vane control) will vary with water flow rate Q and how variable speed operation of the turbine rotor can track the maximum available power.

The paper describes the application of permanent magnet synchronous generator (directly driven by the turbine, i.e. with no gearbox) with a full converter to grid-connected micro-hydro operation (less than 1 kW). The converter combines a three-phase diode bridge, a DC–DC boost converter, a DC link and a three-level

Figure 4.52 Use of variable speed drive to track changes in water flow rate. Source: Courtesy of Marquez et al. [14].

Figure 4.53 Variable speed grid-connected mini-hydro arrangement. Source: Courtesy of Sikorski and Korzeniewski [13].

voltage-source inverter. It is connected to the network via a passive line filter and step-up coupling transformer. An appropriate converter control system is described and demonstrated to effectively track changes in water flow rate, while maintaining unity power factor.

Just as with wind, other configurations are possible. Reference [13] describes a variable speed system using an induction generator with gearbox for mini-hydro systems in the range of 20–500 kW. The motivation was to overcome the limitations of direct connected induction generators for these hydro applications; namely: high start-up currents, lack of possible generation at low water levels and adverse impact on grid power quality. The converter combines a controlled rectifier, DC link and three-level three-phase inverter that can provide power factor control. Figure 4.53, taken from Reference [13], shows the overall arrangement and the control system. The paper also shows that, with suitable changes to the power electronics control, the system can be operated in a stand-alone, non-grid-connected mode.

More complex systems for stand-alone operation have also been suggested, for example using a DFIG machine to supply the local loads, as shown in Figure 4.54, taken from Reference [15]. A DFIG was chosen in this case to reduce cost, say compared to a PMSM, although it should be noted that a gearbox is required in this arrangement. In either case, though, adding a further machine would seem unnecessary, since as outlined before, the inverter can be controlled to supply these loads directly, in fact by operating as a controlled voltage source.

4.8 Applications to Tidal Stream Turbines

Tidal stream turbines resemble wind turbines and they are also subject to flow turbulence, so it might be expected that they would adopt the same variable-speed drives as used by the wind sector, and to date they have done so. There is, however, one key issue

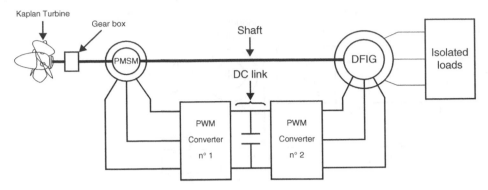

Figure 4.54 Alternative micro-hydro stand-alone configuration. Source: taken from [15].

that is presently gaining attention in this application context, and that is the lack of reliability of power electronic components. Power electronic devices and control cards can be relatively easily changed in wind turbines, even if this means offshore access, but tidal stream turbines are located sub-sea where access is much more challenging and expensive. System designers are considering the implications of moving the power electronic converters to the shore. They are then rather distant from the electrical machines and this raises a number of technical challenges. Similar challenges are encountered with bore hole pumps for oil extraction, where cable lengths can be up to tens of kilometres. Care is needed to avoid excessive voltages, electromagnetic travelling waves in the cables and system resonance that can arise from an interaction between the inverter and the cable impedance. This is not a new issue and research dates back at least 15 years, see, for example, Reference [16]. Recent papers specifically addressing the issues of onshore converters for tidal stream arrays and proposing solutions to these problems include References [17, 18].

References

1 Smith, R.J. and Dorf, R.C. (2007). *Circuits Devices and Systems*, 5e. Wiley.
2 Gross, C.A. (1986). *Power System Analysis*. Wiley.
3 Lander, C.W. (1987). *Power Electronics*. McGraw Hill.
4 Slootweg, H. and De Vries, E. (2003). Inside wind turbines: Fixed vs. variable speed. *Renewable Energy World* 6: 1, 31–40.
5 Pedersen T.K., 'Semi-variable Speed operation – a Compromise?' Vestas-Danish Wind technology A/S, Denmark, EWEA 1995 European Conference, 1995.
6 Kruger T. and Andersen B., Vestas Optispeed – advanced control strategy for variable speed wind turbines. Department for Research and Development, Vestas Wind Systems A/S, Jacobsens Vej 7, DK-6950 Ringkobing, 2011.
7 Ekanayake, J., Holsworth, L., and Jenkins, N. (2003). Control of DFIG wind turbines. *Power Engineer* 17 (1): 28–32.
8 Blaabjerg, F., Liserre, M., and Ma, K. (2012). Power electronic converters for wind turbine systems. *IEEE Trans. Ind. Applications* 48 (2): 708–719.

9 Ackerman, T. (2002). Transmission systems for off-shore wind farms. *Renewable Energy World*: 49–61.

10 Wind Denmark. The Danish Wind Industry Association website https://winddenmark.dk/en (accessed May 2019).

11 Heier, S. (2006). *Grid Integration of Wind Energy Conversion Systems*, 2e. Wiley.

12 Jamieson, P. (2011). *Innovation in Wind Turbine Design*. Wiley.

13 Sikorski, A. and Korzeniewski, M. (2011). AC/DC/AC converter in a small hydroelectric power plant. *Bull. Pol. Acad. Sci. Tech. Sci.* 59 (4): 507–511.

14 Marquez, J.L., Molina, M.G., and Pacas, J.M. (2010). Dynamic modelling, simulation and control design of an advanced micro-hydro power plant for distributed generation applications. *Int. J. Hydrog. Energy* 35: 5772–5777.

15 Ansel, A. and Robyns, B. (2006). Modelling and simulation of an autonomous variable speed micro hydropower station. *Math. Comput. Simul.* 71: 320–332.

16 Smith, J.R., Grant, D.M., Al-Mashgari, A., and Slater, R.D. (2000). Operation of subsea electrical submersible pumps supplied over extended length cable systems. *IEEE Proc. Electr. Power Appl.* 147 (6): 554–552.

17 Sousounis, M.S., Shek, J.K.H., and Mueller, M.A. (2016). Modelling, control and frequency domain analysis of a tidal current conversion system with onshore converters. *IET Renew. Power Gener.* 10 (2): 158–165.

18 Sousounis, M.S., Shek, J.K.H., and Mueller, M.A. (2016). Filter design for cable overvoltage and power loss minimization in a tidal energy system with onshore converters. *IEEE Trans. Sust. Energy* 7 (1): 400–408.

5

Power-System Analysis

5.1 Introduction

The preceding chapters have discussed electrical generators in some detail. Attention will now turn to the transmission and distribution systems that carry the generated power to the loads. In addition to their obvious task of carrying electricity over a distance and allowing generators and loads to be geographically separated, transmission and distribution systems also serve to provide interconnection. The interconnection of loads provides valuable aggregation and hence smoothing, as discussed in Chapter 3. Also, the parallel operation of generators allows inefficient part-loaded operation of individual generators to be largely avoided (Chapter 7) and, thus, overall efficiency is greatly increased. Furthermore, the parallel operation of generators and of sections within the transmission and distribution systems provides redundancy, allowing the overall impact of any single component failure to be minimised. Thus, even in situations where generators are located close to loads, the benefits of an interconnected power system should not be underestimated.

The effective operation of a power system requires that the routing of power from generators to consumers is carried out through the network so that transmission equipment is not overloaded, losses in the network are kept to a suitably low level, voltages at various nodes are maintained within limits, and loss of generation or transmission equipment does not lead to an unacceptably high frequency of black-outs. To ensure that power systems operate securely and efficiently, special computational techniques have been developed by utilities. This chapter briefly describes these techniques and illustrates their use in determining the voltage at the nodes of the network, the complex power flows in transmission lines and the currents during fault conditions. The effect on the network of adding renewable energy (RE) generators can be assessed using such techniques, as will be examined in Chapter 6.

Finally, the issue of power-system stability will be discussed in outline. This is a complex and challenging subject and a detailed analysis is beyond the scope of this book.

5.2 The Transmission System

Figure 1.18 in Chapter 1 gave a pictorial view of a typical power system and the desirability of using various voltage levels for transmission were explained. Overhead lines, supported on steel towers (pylons), concrete towers or wooden poles, are highly visible

Renewable Energy in Power Systems, Second Edition. David Infield and Leon Freris.
© 2020 John Wiley & Sons Ltd. Published 2020 by John Wiley & Sons Ltd.
Companion website: www.wiley.com/go/infield-renewable-energy

33 kV   132 kV   11   400 kV

Figure 5.1 Overhead lines at four different voltage levels. Source: Courtesy of Murray Thomson.

in most power systems worldwide. Figure 5.1 shows the typical appearances of overhead lines at various voltage levels in the UK. It should be noted, however, that pole and tower constructions vary considerably and do not always provide a clear indication of the voltage level in use. Note also that underground cables, despite being much more expensive to install, are widely used at the lower voltage levels to avoid overhead transmission line intrusion in areas of habitation [1].

The 11-kV overhead line on wooden poles in Figure 5.1 is three-phase with no neutral. This is called a *single circuit*. The steel tower lines are carrying *two* three-phase circuits. In the 400-kV line each of the six conductors is made up of a bundle of four individual conductors. Each bundle behaves effectively as a single conductor, but the spacing helps to reduce the line impedance and the losses and increases the line capacity slightly. The two three-phase circuits can be operated independently, which provides operational flexibility and greatly increases reliability. Such an arrangement is known as a *double circuit*. Again, there is no neutral: the wire visible at the very top of the tower is an earth wire for lightning protection.

The voltages indicated in Figure 5.1 are widely used in the UK. Other countries use slightly different voltages, often with fewer intermediate levels, but the structure is always similar. The terms *low-voltage* (LV), *medium-voltage* (MV), *high-voltage* (HV) and *extra-high-voltage* (EHV) are often used but are avoided here because definitions vary considerably from country to country.

Transformers, located in substations (Figure 5.2) or on poles, are used to transfer power between lines and cables operating at different nominal voltages. A mature network may have tens of thousands of transformers. Fortunately, transformers are both very reliable and efficient.

Last, and by no means least, a great deal of switchgear is installed as shown in Figure 5.2. This allows sections of the system to be isolated for operational and maintenance purposes, or on the occurrence of faults.

Figure 5.2 Transformer: 132–133 kV. Source: Courtesy of Murray Thomson.

5.2.1 Single-Phase Representation

Appendix A explains in detail the reason why three-phase is exclusively used for the generation and transmission of electrical power. It is also shown that as long as the three phases are balanced, calculations can be performed by considering just one of the phases. Thus, it is common to draw just one phase in circuit diagrams. To achieve this, delta-connected components are represented by their equivalent star and a neutral conductor may be drawn in, even when there is no neutral in reality.

To simplify drawings even further, it is common to omit the neutral return and represent a three-phase generator supplying a load as in the *one-line diagram* of Figure 5.3.

This one-line short hand is universally used by power-system engineers both to represent schematically and perform calculations in balanced power systems.

A one-line diagram of a power system is shown in Figure 5.4. The different voltage levels found in a large power system may be thought of as layers, each one feeding the one below it by way of transformers.

5.2.2 Transmission and Distribution Systems

It is usual to refer to the higher-voltage levels as the *transmission system* and the lower voltage levels as the *distribution system*, but the exact usage varies from country to country. In England and Wales, the 400 kV (plus some sections at 275 kV) is regarded as the transmission system, while the 132 kV and all layers below are the distribution system. In Scotland, many sections of 132 kV are regarded as transmission.

Figure 5.3 One-line diagram.

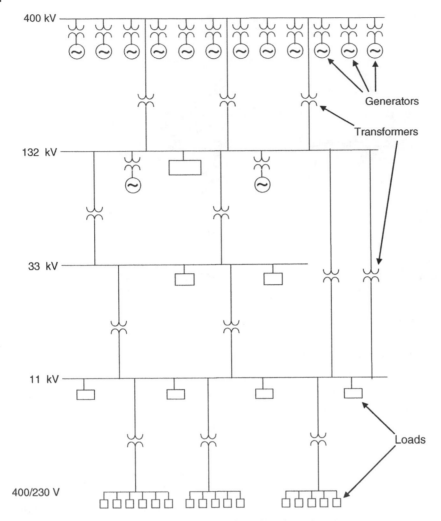

Figure 5.4 One-line diagram showing 'voltage layers' in a large power system.

The term *national grid* usually means the transmission network. However, the term *grid* is often used loosely to describe the transmission *and* distribution networks. In particular, *grid connected* means connected to any part of the network but is used increasingly to mean small generators connected directly to the low-voltage distribution system.

Nearly all generation, currently installed, is connected to the transmission system and is called *centralised generation*. Thus, traditionally, power always flows to the consumers *down* through the distribution system. And, traditionally, distribution systems are designed with this in mind.

5.2.3 Example Networks

The transmission system shown in Figure 5.5 covers the length and breadth of the UK mainland. It is a single AC power system and therefore has the same frequency

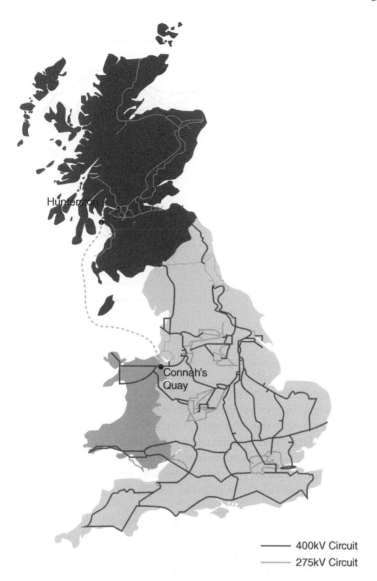

Figure 5.5 UK transmission system (also showing the planned HVDC offshore interconnector). Source: Image from Scottish Power.

throughout, but will soon include the shown offshore high-voltage direct current (HVDC) link to improve north–south power flow.

There is a similar, but smaller, transmission system that covers Ireland, including Northern Ireland. It also operates at 50 Hz nominal, but the frequency is not tied to the UK mainland system. There is a HVDC link from Scotland to Northern Ireland that allows power to be transferred in either direction, but does not require that the two systems are run in synchronism. There is a similar, but higher capacity, HVDC link between England and France. France is interconnected with its neighbouring

Figure 5.6 North Wales.

countries, so effectively there is a continuous connection between the most northern tip of Scotland and the most southern tip of Italy.

Most major European countries generate electricity approximately equal on average to their national demand. Nonetheless, interconnections with neighbouring countries are very valuable for efficient operation. Of particular note is Denmark, where the impressive penetration of wind power is facilitated by such interconnections.

Returning to the British Isles, many of the islands around the coast are now connected by under-sea cables and enjoy electricity with a cost and reliability similar to that on the mainland. More remote islands have stand-alone systems, usually with a heavy dependence on diesel fuel.

Figure 5.6 shows in greater magnification the north Wales network that includes only the upper voltage layers. The 400-kV transmission network matches (roughly) that shown in Figure 5.5. The grey blobs in Figure 5.6 are the bulk supply points (transformer substations), where power is fed down to the 132-kV network, which in turn takes power to major towns. Further transformer substations feed power down to the 33-kV network, which takes power to the smaller towns. Figure 5.7 shows the much denser 11-kV network layer that takes power to villages, hamlets, and even individual farms. Finally, the even denser 400/230-V network (not illustrated) distributes power to individual properties as shown schematically in Figure 1.18.

Further information regarding the UK power system can be found in Reference [1].

5.3 Voltage Control

Consumer loads comprise mainly: heating and lighting elements, motors, electronic equipment providing audio-video services, computers and controllers, battery charging

Figure 5.7 The 11 kV network in South Wales on the same scale.

facilities for portable equipment and finally electro-chemical services mostly for industrial applications. All these 'loads' are designed to provide their service from a nominally fixed voltage supply. Utilities are therefore obliged by law to provide electricity at consumer terminals at a voltage that does not deviate from the nominal value by more than a few percent with a maximum of ±10% not being uncommon. For this to be achieved at the extremities of the distribution network, the voltage of all nodes moving up through the layers of increasing voltage in the distribution and transmission system should be kept close to their nominal value. As the power-system engineers would put it 'the voltage profile from the generator terminals down to the consumer terminals should be kept nearly flat'.

The maintenance of a nearly flat voltage profile is disturbed by the fact that current flow through the impedance of transmission lines, cables, and transformers causes unavoidable voltage drops. For these drops to be kept within permissible limits, a number of *compensation* techniques have been developed over the years. These techniques have been considerably extended recently through the use of power electronic devices capable of high-voltage and current ratings. Before discussing these techniques it is necessary to examine the characteristics of transmissions lines and the effect on voltage drops of the active and reactive power flows over the network.

5.4 Power Flow in an Individual Section of Line

Calculations to quantify power flows and exact node voltages throughout a network must take into account the network impedances. Later, it will be shown that because of the operational complexities of power systems these calculations are not trivial and require sophisticated software for their execution. To gain an appreciation of these calculations and the relationships between power flows and network voltages, it is necessary first to consider a single point-to-point section of transmission or distribution line.

5.4.1 Electrical Characteristics of Lines and Cables

The conductors used in overhead lines and underground cables are almost universally copper or aluminium, having a small but finite series resistance. They also have self and mutual inductance and shunt capacitance to each other and to ground (not shown), as illustrated in Figure 5.8.

The resistance, inductance and capacitance are distributed along the length of the line and are quantified per-unit-length. For most calculations, however, they can be considered as lumped components with values calculated for the length of the line in question.

5.4.2 Single-Phase Equivalent Circuit

Assuming that its operation is balanced, the three-phase line shown in Figure 5.8 can be reduced to a single-phase equivalent as shown in Figure 5.9. The capacitance has been removed in order to simplify the following calculations. Its effect is often negligible for short overhead lines. However, for long HV lines and invariably for underground cables even of short length, the capacitance is substantial. For such cases, the circuit of Figure 5.9 can still be used for the first stage of the analysis. The capacitance can be lumped at the two ends of the line and its effect taken into account at a second stage in the calculation.

In Figure 5.9, the upper line (containing the impedance) is one of the three lines shown previously in Figure 5.8. The lower line is a neutral, drawn in to assist visualisation even

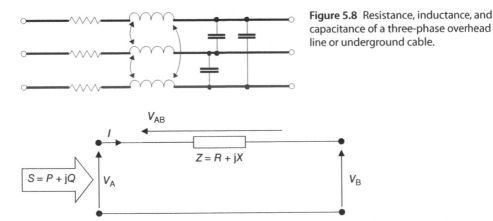

Figure 5.8 Resistance, inductance, and capacitance of a three-phase overhead line or underground cable.

Figure 5.9 Single-phase equivalent circuit of a transmission line.

though there is no neutral in reality. It is not known whether the components at the ends of the line are star or delta connected, but the calculations are being done as if both were star and a neutral were present.

P is the active (real) power flowing into the left end of the line, and Q is the associated reactive power, defined in the same direction. They are jointly represented by the complex power S:

$$S = P + jQ \qquad (5.1)$$

In this case, S, P, and Q are per-phase quantities and will be one third of the overall values carried by the real three-phase line.

V_A is the complex phase voltage (line-to-neutral) at the left end of the line. It is equal to the line voltage (line-to-line) divided by $\sqrt{3}$.

I is the complex line current and is the same as the phase current in this context.

The complex power is related to the voltage and current by (Appendix A):

$$S = V_A I^* \qquad (5.2)$$

where I^* is the complex conjugate of I.

5.4.3 Voltage Drop Calculation

In Section 4.2.5, the synchronous machine equivalent circuit consisting of a single series reactance was used to express the relationships between the active and reactive power flows as functions of magnitudes and angles of the equivalent circuit terminal voltages (note the similarity between Figures 5.9 and 4.7). In what follows, the inverse relationship will be derived because there is often interest in the way the voltage profile of the network from node to node is affected by the active and reactive power flows.

The complex impedance Z shown in Figure 5.9 includes resistance R and reactance X. The resistance is the same as that shown previously in Figure 5.8. The reactance is calculated from the inductance shown in Figure 5.8 plus mutual inductance terms. Such calculations are beyond the terms of reference of this book and can be found in any textbook on power systems; for example, Reference [2].

There will be a voltage across this impedance, according to Ohm's law:

$$V_{AB} = IZ \qquad (5.3)$$

And, by Kirchoff's law:

$$V_B = V_A - V_{AB} \qquad (5.4)$$

The phasor diagram of Figure 5.10 represents pictorially all the variables of Figure 5.9. Here, the current I at end A is shown lagging the voltage V_A by about 30°, hence the transmission line at A is a sink of active ($+P$) and reactive ($+Q$) power and $S = P + jQ$.

Figure 5.10 Phasor diagram illustrating voltages and current shown in Figure 5.9.

Combining Eqs. (5.2) and (5.3), and taking V_A as the reference phasor with an angle of zero, gives:

$$
\begin{aligned}
V_{AB} &= \frac{S^* Z}{V_A} \\
&= \frac{(P - jQ)(R + jX)}{V_A} \\
&= \frac{PR + QX}{V_A} + j\frac{PX - QR}{V_A}
\end{aligned}
\tag{5.5}
$$

5.4.4 Simplifications and Conclusions

The use of complex numbers, as exemplified before, provides a systematic and accurate means of calculating magnitudes and phase angles of all quantities. Good engineering calculators and many software packages can manipulate complex numbers directly and make such calculations straightforward.

Nonetheless, the following approximations to Eq. (5.5) are often very helpful, both for rough calculations and to get a better feel for the interdependencies.

First, there is often interest in the scalar difference between the magnitudes of voltages V_A and V_B. Referring to Figure 5.10, it can be seen that this difference is contributed mainly by the real part of V_{AB} (its horizontal component). Therefore, Eq. (5.5) can be simplified to give a scalar relationship:

$$
\Delta V \approx \frac{PR + QX}{V}
\tag{5.6}
$$

where $\Delta V = |V_A| - |V_B|$. The denominator V should really be $|V_A|$, but $|V_B|$ can be used, provided that ΔV is small.

Equation (5.6) provides a convenient way to estimate the voltage at one end of the line compared to the other, for a given active and reactive power transfer over the line.

Moreover, the same equation can be used to estimate the *voltage change* that will occur at the point of common coupling (PCC) when a new generator (or load) is connected to a system. In this case, the R and X need to represent the Thévenin equivalent impedance (Appendix) of the system as a whole, but more on this later.

A second approximation is applicable in transmission networks and the higher layers of distribution where, due to the geometry and size of the conductors, the impendence Z is dominated by the reactance X. Table 5.1 shows typical X/R *ratios* for high to low-voltage overhead transmission lines.

Table 5.1 Transmission line parameters

kV	Typical X/R ratio
400	16
275	10
132	6
33	2
11	1.5

When the X/R ratio is high it can be assumed that $R = 0$ in Eq. (5.6) with little loss in accuracy. This gives the very useful rule of thumb:

$$\Delta V \propto Q \tag{5.7}$$

In Figure 5.10 it can be seen that the angular displacement between V_A and V_B is to a large extent proportional to the imaginary component of the V_{AB} phasor. The imaginary part of Eq. (5.5) gives a second rule of thumb:

$$\delta \propto P \tag{5.8}$$

where δ is the load angle between V_A and V_B, the same angle that was encountered in Chapter 4 when the operation of the synchronous generator was under study. Equations (5.8) and (5.7) match Eqs. (4.8c) and (4.8d).

These two relationships provide an important insight in the flow of complex power in transmission networks where $X \gg R$:

- Network voltages are determined largely by reactive power flow. Turning round this statement it can be affirmed that voltage *magnitude differences between the ends of a transmission line is the primary driver of reactive power flow.*
- Phase angles are determined largely by active power flow and are largely independent of reactive power flows. Turning round this statement as well it can be affirmed that *angular differences between the voltages at adjacent nodes is the primary driver of active power flow.*

These insights can be extended further. Consider that a power-system component, capable of absorbing or injecting active or reactive power, is connected to a node of the transmission network. If the component were to inject or extract active power at the node, the node voltage angle will shift to a more leading or lagging position respectively with respect to the voltages of the adjacent nodes. If the component were to inject or extract reactive power at the node, the node voltage magnitude will increase or decrease respectively compared to its original value. It may be concluded that if a flat network voltage profile is to be maintained, the high-voltage can be depressed or low-voltage augmented at a node by connecting inductors or capacitors respectively to that node.

5.5 Reactive Power Management

The principle of conservation of reactive power described in Section A.13, requires that reactive power is scheduled so that as the consumer demand of Q varies over a day, appropriate Q resources are available to provide it. In Chapter 3, it was explained that a discrepancy between demand and supply of active power is manifested by rises or falls in frequency, a parameter that is common to the *whole* interconnected power system. In contrast, reactive power deficits or excesses manifest themselves as *local* system voltage drops or rises respectively. An overall deficit of reactive power in a system will manifest itself as unacceptably drooping voltage profiles. The reverse is the case for a surfeit of reactive power.

Ideally, reactive power should not be transported but generated locally where it is required. This, of course, is the principle of power-factor correction of consumer loads, an arrangement described in Appendix A, Section A.14 and encouraged by utilities through tariffs. However, the vast majority of domestic and commercial loads as well as some industrial loads, are not power factor corrected. Hence, substantial amounts of reactive power have to be transported over the system and/or generated locally by special utility equipment. The picture is complicated by the fact that:

a) High-voltage overhead transmission lines overall generate Q when they are lightly loaded due to their shunt capacitance but absorb Q when heavily loaded due their series inductive reactance (Figure 5.8);
b) Low-voltage transmission lines have small shunt capacitance and overall are consumers of Q;
c) Cables overall generate Q at all loads because of their substantial shunt capacitance;
d) Transformers always absorb Q as they consist basically of two inductively coupled coils (Section 4.4.2).

5.5.1 Reactive Power Compensation Equipment

Utilities use the following devices to balance the Q flows in the network and therefore control the bus voltages:

1. transformer tap changers
2. synchronous generator Automatic Voltage Regulators (AVRs)
3. shunt capacitors
4. shunt reactors
5. static VAR Compensators (SVCs)
6. fast AC Transmission Systems (FACTS)
7. renewable energy generator interfaces.

5.5.1.1 Tap Changers and Voltage Regulators

Tap changers play a major role in voltage control. A range of connection points (taps) to the windings of a transformer allows the turns-ratio to be altered and thus provides voltage control. An on-load tap changer is a collection of motorised switches allowing adjustment while the transformer is in service.

Tap changers do not themselves enable the generation or consumption of Q. Their function is to regulate the reactive power flow over the network from generators to consumers. An example of how this is done is described in the next sub-section.

Figure 5.11 shows that on-load tap changers (transformer symbol with an arrow) are widely used throughout power systems, except on the 11 kV to 400 V final distribution transformers. There are a great many distribution transformers and the maintenance of motorised tap changers would be impractical. Instead, the taps on a distribution transformer can be adjusted manually, but this requires taking the transformer out of service.

Thus, delivery of the correct voltage to domestic customers (230 V ±10%) relies heavily on the tap changer at the primary substation that feeds the 11 kV network. Because of its crucial importance this tap changer is activated through an *AVR*. The allowable voltage range, ±10%, may seem generous, but in practice it is often fully exploited to allow for load variation between summer and winter.

Figure 5.11 Voltage levels and interconnections to regulate Q flow in a typical transmission network.

5.5.1.2 AVRs

The internal source voltage in the stator of a synchronous generator (Figure 4.7) can be controlled by controlling the DC current (the excitation) in the field winding on the rotor. This control is often achieved by an AVR. This is an electronic controller that monitors the voltage at the synchronous generator terminals, compares it to a set value and if there is an error increases or decreases the excitation to nullify it. In the case of a stand-alone synchronous generator, the AVR's role is straightforward: it monitors and controls the terminal voltage.

Large conventionally fuelled generator units are connected to the grid through a 'unit' transformer fitted with a tap changer. Most auxiliaries that supply services to the generator are connected to the generator terminals and require a fixed voltage to provide the appropriate service. The task of the AVR here is to maintain the generator terminal voltage constant.

The level of reactive power injected to the grid by a generator connected to the grid through a transmission line is controlled by the unit transformer tap changer as follows. By increasing the nominal transformer ratio of say, 22:400 kV, by tapping-up, the transformer secondary voltage increases with respect to the remote end voltage of the transmission line and more reactive power is transported down the line. This extra reactive power flow through the synchronous reactance of the generator results in a reduction of its terminal voltage. This is sensed by the AVR, which increases the excitation of the generator (hence, the injection of extra reactive power) to bring the voltage up to the nominal level.

Generators fed from RE sources are substantially smaller in rating and are, in general, connected to the distribution rather than the transmission network. The effect on the system of such generators are discussed in Chapter 6. However, if the RE source is truly large, for example a large offshore wind farm or a tidal scheme, it could have an impact on the network. In such cases, more conventional control schemes may have to be used. In fact, as penetration levels by wind power is increasing; for example, in Denmark, where levels of more than 40% are no longer uncommon, Utilities require that renewable generators are designed to contribute positively to power system voltage control.

5.5.1.3 Static Compensators

At the transmission network, switchable capacitors and/or inductors are connected at crucial nodes of the system to regulate the voltage under varying loading conditions. These devices are controlled in steps and in the case of inductors are particularly lossy. For finer control, SVCs are used. These employ solid state power electronic devices rather than mechanical switches to provide infinitely variable control of Q from reactors and capacitors.

5.5.1.4 FACTS

The way pulse width modulation (PWM) converters can be controlled to act as capacitors or inductors was discussed in Section 4.5.6. FACTS are power electronic converters that can be controlled to adjust the angle and voltage at a network location. In this way, they can be designed to transfer no active power at all and act as pure reactive power sources or sinks.

5.5.1.5 RE Generator Interfaces

In Chapter 4 it was seen that to maximise effectiveness or because of absolute necessity the majority of renewable energy generators are interfaced to the grid through a power conditioning converter. This has the added advantage that if the converter is of the PWM type it can contribute to the voltage regulation at the PCC.

5.6 Load-Flow and Power-System Simulation

5.6.1 Uses of Load Flow

Section 5.4 looked at the mathematics describing an individual overhead line or underground cable. Mature power systems are likely to have thousands of such lines, all interconnected. The same basic mathematics applies to each and every line, but now the equations must be solved simultaneously. Structured procedures for such calculations are known as *load-flow* [2, 3].

A basic load-flow calculation provides information about the voltages and currents and complex power flows throughout a network, at a particular point in time, with a given set of load and generation conditions. Additional information, such as losses or line loadings, can then be easily calculated.

Load-flow analysis is an essential tool that provides the following vital information for the design as well as the operation and control of power systems:

- Checking whether equipment run within their rated capacity,
- Checking that voltages throughout the network are kept within acceptable limits,

- Ensuring that the power system is run as efficiently as possible,
- Ensuring that the protection system will act appropriately under fault conditions and that under likely contingencies the system will remain secure and operational,
- Assisting in the planning of the expansion of conventional and renewable generation and the necessary strengthening of the transmission and distribution system to meet future increases in power demand.

Load-flow calculations are central to the numerous software packages available for power-system simulation, including PSS/E and DIgSILENT. Most such packages also include many additional functions, such as the following:

Fault analysis determines the currents that will flow in the event of a short circuit. Such information is critical in the design of switchgear and protection systems.

Unbalanced analysis allows modelling of networks where the three phases are not perfectly balanced. This is particularly relevant in networks where single-phase loads are connected.

Dynamic simulation allows transients and stability to be examined. This is particularly important in the design of transmission systems. The modelling has to include the dynamic characteristics of the generators and their associated control systems.

5.6.2 A Particular Case

The one-line diagram of the Icelandic national power system shown in Figure 5.12; the system is powered mostly by renewable energy, primarily hydro, but with an increasing contribution from geothermal.

Figure 5.12 Single-line diagram of the Icelandic national power system. Source: Reproduced from Egill B. Hreinsson http://www.hi.is/~egill/rit/estonia99.pdf.

The generators are indicated by the ⊙ symbol, while the loads are shown as triangles. In practice, these loads are the bulk supply substations that feed power to the distribution networks, but for now they will be thought of as simple point loads. The thick horizontal lines are *nodes* (also known as *buses*) and, although they have a finite length in the diagram, *electrically* they are just connection points: they have zero length and zero impedance. The thinner lines in the diagram are the actual transmission lines and each one has a finite impedance, which may be estimated from the line characteristics and its length. The system uses a mix of 220, 132 and 66 kV, and the transformers connecting these are indicated by the usual ⊖ symbol.

In the design and operation of a power system, it is important to know the power flows, both active and reactive, the voltages (which will normally differ slightly from the nominal values mentioned previously) and the currents *throughout the system*. The almost illegible numbers in Figure 5.12 are the values of these variables, which have been calculated by load-flow analysis. How such calculations are performed will now be considered.

Consider the transmission line running across the middle of the diagram from node *Brennimelur* to node *Hrauneyjarfoss*. The power transfer over this transmission line will obey the mathematical analysis presented earlier in this unit. Indeed, the power transfer over every individual transmission line in the whole system must obey the equations presented earlier. The challenge is that, in general, the equations require to be solved *simultaneously*.

While the mathematical techniques and the software packages for performing load-flow analysis are readily available, obtaining suitable input data can be much more difficult. The required input data can be divided into network data and load/generation data.

5.6.3 Network Data

The impedances of the individual lines are usually calculated from knowledge of the line type and the length of the line. In practice, particularly in lower voltage networks, these are not always known with absolute certainly. The line may have been installed fifty years ago and could have been modified since. Utilities do not always have perfect records. Also, particularly with underground cables, the details of the installation can significantly affect the characteristics; for example, the dampness of the ground and the proximity of other cables can have a significant effect.

The ratios and impedances of the transformers must also be known and again, records can lack details. Any tap-changing mechanisms on transformers must also be included in the model.

5.6.4 Load/Generation Data

5.6.4.1 Time Dependence
One load-flow analysis solution provides the results for one operational instant. If any of the loads or generator outputs are changed, then the analysis must be run again. With modern computers, rerunning the analysis is trivial; the challenge is in deciding what data should be used for loads and generators, given that they are continuously varying.

A common approach is to identify the worst cases, and typical choices are the winter maximum peak demand and the summer minimum, except in the US where the maximum is due to summer air-conditioning. At the winter peak, the system is stretched to its limit and it is necessary to ensure that no equipment is overloaded or any node voltage below the minimum permitted. During the summer minimum, because of the low loading of transmission lines and cables, there may be an excessive reactive power available that would manifest itself as high-voltage at nodes.

In a transmission system, the loads and generation normally vary smoothly and the relevant maximum and minimum figures can readily be identified. In lower voltage networks, the loads are far more variable and the worst case conditions can be very difficult to identify.

5.6.4.2 Types of Nodes (Buses)

PQ Buses The obvious way of describing the load at a node is by specifying it as an impedance. Simple uncontrolled loads, such as heaters and incandescent light bulbs, tend to behave as near-constant impedances and, thus, the power they draw varies with the square of the voltage: a $\pm 10\%$ variation in voltage, which is not uncommon, gives powers ranging from 81 to 121%. Many loads, however, include some sort of control mechanism such that the power they draw does remain near constant irrespective of the voltage changes. Most electronic-based devices behave in this way.

Another important example is a distribution network that is supplied via a tap-change transformer. Such a transformer senses the voltage on the lower voltage side and changes the taps so that this is maintained at nearly nominal value. A fixed load impedance on the transformer secondary appears on the primary side as a fixed P and Q demand irrespective of the transformer ratio. As the taps change the voltage on the primary side of the transformer may change substantially but the P and Q are invariant. In Figure 5.11 the tap-change transformer at the primary substation can be adjusted to achieve a constant voltage on the 11 kV side irrespective of (modest) voltage changes on the 33 kV side. Thus, the power drawn by the 11 kV network from the 33 kV network is also held near constant. If one were performing a load-flow calculation on the 33 kV network, which would typically feed several 11 kV networks, it would be reasonable to represent each primary substation as a constant P and Q. The same is true in load-flow modelling of the higher-voltage networks.

To summarise, assuming that the voltage at the bus is kept constant the load can be expressed as a fixed P and Q demand and such nodes are referred to as *PQ buses*. As will be explained later, this is a very convenient formulation for the purposes of carrying out the load-flow.

Small renewable energy generators also fall in the PQ bus category. Distributed generators are infrequently called upon to control the network voltage. Instead, they are often configured to operate at near unity power-factor ($Q = 0$); in which case, it may be appropriate to label the node to which they are connected as a *PQ node*. In the case of fixed-speed wind turbines, however, the reactive power consumed by the induction generator will be dependent on voltage, as will the reactive power generated by the power-factor correction capacitors. Many load-flow software packages include facilities to appropriately model induction machines and related equipment. The situation is similar with small hydro systems interfaced to the grid through induction generators.

Energy from photovoltaic, wave and tidal schemes and MW size wind turbines is fed to the grid through a power electronic converter. This provides the facility of reactive power injection/extraction at the point of connection.

To summarise, for relatively small embedded RE generators, the P injection depends solely on the RE source (wind, sun, water) level at the time and the Q injection either on the bus voltage or on the setting of the power electronic converter. In the latter case, the converter could be regulated to inject active power at a chosen power factor.

According to the generally accepted convention summarised in Figure A.20, at PQ nodes generators inject active power and so P is positive, whereas for loads, P is negative. The reactive power Q direction is defined similarly.

PV Buses   For large synchronous generators, the Q is often not specified and, instead, it is the voltage V that is known. This is because such generators are fitted with AVR's which hold the V constant. To accommodate this in load-flow analysis nodes where such generators are connected are referred to as *PV buses* and are dealt with a little differently in the mathematics. Unfortunately, PV buses are sometimes described as *generator buses*, which makes sense so long as all the generators are large synchronous generators with AVRs.

As renewable energy generators increase in size, Utilities have been developing regulations requiring that such generators behave like traditional generators. So, multi-megawatt wind turbines connected to the network through PWM inverters may be required to regulate the local bus voltage. In such cases, the node has to be treated as a PV bus.

Slack, Swing or Reference Bus   The two types of buses described earlier require that the Ps are specified at all the network nodes before the load flow calculations are initiated. This does not make sense in terms of the conservation of power principle because the transmission losses in the system are not known before the solution is arrived at! This conundrum is resolved by allowing one bus that has a generator connected to it to be specified in terms of the magnitude V and angle δ of its voltage. Such a bus is known as *slack, swing* or *reference bus*. The voltage at this bus acts as the reference with respect to which all other bus voltages are expressed. At the end of the load flow, the calculated P and Q at this bus take up all the *slack* associated with the losses in the transmission.

5.6.5   The Load-Flow Calculations

In small networks, it is often possible to obtain valid and useful results by direct application of the mathematical analysis presented earlier. Also, larger networks can often be *reduced* to equivalent circuits that can be solved in the same way. However, load-flow analysis of any system beyond a few nodes is carried out by a computer.

Each bus of a power network is characterised by a number of variables. For a network with predominantly reactive transmission line impedances all these variables are linked by the complex Eq. (5.9), replicated here.

$$S_B = -\frac{V_A V_B}{X_t}\sin\delta + j\left(\frac{V_B}{X_t}(V_B - V_A\cos\delta)\right) = P_B + jQ_B \tag{5.9}$$

This equation describes the performance of a synchronous generator but it is equally applicable to a transmission line linking two buses. Note that there are four variables associated with each bus. The active and reactive power injected or extracted at the bus and the magnitude and angle of the bus voltage. The complex Eq. (5.9) can be split into two separate equations one for the real and another for the imaginary part. For a network with n nodes, there are $2n$ simultaneous equations to be solved, hence at each node two of the four variables have to be specified. All of this fits neatly with the earlier arrangement of specifying buses in terms of P and Q, P and V or V and δ.

The solution of the $2n$ equations describing the network is not a trivial task. The basic Eq. (5.9) linking the node variables is non-linear because it contains products of the variables and trigonometric functions. The load-flow solution to such non-linear simultaneous equations require iterative techniques and are beyond the terms of reference of this book. In general, therefore, the solution of most practical networks requires a very structured approach to the mathematics and are carried out by computers. Indeed, it was one of the first useful applications of computers during the 1950s. There are many mathematical techniques for performing load-flow analysis. They are all iterative, and most begin by representing the network as a matrix. They often use Newton–Raphson, Gauss–Seidel iterative methods or embellished versions of these. They should all converge to very similar answers and the choice between the methods is mostly a matter of speed and reliability of convergence. The different methods can perform better or worse with different types of networks. For more information, consult the power-system simulation packages PSS/E, DIgSILENT. A recent trend is that load-flow software is being integrated into geographical information systems (GIS) and this integrates with the asset management side of the electricity utility business.

The assumption of constant voltage at the PQ buses inherent in the formulation of the load flow described before is not valid at the distribution level. The distribution transformers (11 kV to 400 V) shown in Figure 5.11 are not controlled and representing them as constant P and Q will lead to some inaccuracy. Load-flow software that is intended for modelling the lower voltage levels of distribution networks often includes the facility to define loads that do vary with respect to voltage. However, obtaining load data that includes this characteristic may be less straightforward.

5.6.6 Results

The basic results provided by a load-flow analysis are the power flows (both active and reactive) and the voltages and currents (both magnitude and phase) for all the lines and nodes throughout the system (strictly speaking, some of this information is redundant). From these basic results, an engineer can readily identify or calculate:

- Any overloaded lines or transformers;
- Any voltages that are under or over acceptable limits;
- Power losses, both active and reactive, for individual lines or for the whole system.

This information influences the design and operation of the network. Load-flow studies are routinely performed to check the state of the network as the demand varies through the day or whenever changes to the network or to generation are being considered.

5.6.7 Unbalanced Load-Flow

Ideally, the voltages and currents in three-phase power systems are perfectly balanced, which leads to efficient operation and greatly simplifies any analysis. However, in low-voltage networks particularly, the phases are not always balanced in practice and full analysis requires *unbalanced* load-flow techniques. Unbalanced load-flow usually employs the method of *symmetrical components*, which involves the description of the unbalanced network as the sum of three simpler networks. Many software packages can perform unbalanced load-flow analysis (perhaps sacrificing other features) but in practice the data they require is often unavailable.

5.7 Faults and Protection

Faults in power systems can be caused by a whole range of events. Lightning strikes, overhead lines breaking in storms and workmen digging through cables are common examples. Faults often result in the conductors being effectively shorted together or shorted to ground. In either case, very large currents can flow. The system must be designed to withstand these currents for short periods and to disconnect the faulted section quickly, so that the rest of system continues to operate with a minimum of disturbance. Faults occur quite frequently (very frequently during storms) and must be effectively managed in order to achieve good reliability for the system as a whole. The detection and management of faults in power systems is known as *protection* and is a highly developed aspect of power systems engineering.

At low-power levels, fuses are used and conveniently provide both fault detection and breaking of the circuit. At higher power levels, *protection relays* detect faults and trigger *circuit breakers*. Figure 5.13 shows a typical circuit breaker.

Figure 5.13 A 132 kV circuit breaker. This large and expensive device serves the transformer shown earlier in Figure 5.2. Source: Courtesy of Murray Thomson.

5.7.1 Short-Circuit Fault Currents

The effective management of faults requires knowledge of what currents will flow under all likely fault conditions. First, this knowledge is required in order to design the fault detection mechanism, which has to differentiate clearly between normal load currents (including surges when equipment are first switched on) and fault currents. Second, the magnitude of the possible fault current must be known in order to select the device that has to break the current.

Short-circuit faults can occur in a number of ways, including line-to-ground, line-to-line and so on but, for simplicity, it is usual to start by considering the worst case of symmetrical three-phase faults. Fortunately, this case is also the easiest to analyse.

5.7.2 Symmetrical Three-Phase Fault Current

A symmetrical three-phase fault occurs when all three conductors of a three-phase line are shorted together. This amounts to the same thing as if they were all shorted to ground but, because it is symmetrical (balanced), no current will flow in the ground and a simple single-phase analysis can be used.

In the simple case shown in Figure 5.14, E_g is the internal phase-to-neutral emf of a synchronous generator, X_g is its internal per-phase reactance and R_l and X_l are parameters of the line. By inspection, the fault current I_{sc} is given by:

$$I_{sc} = \frac{E_g}{|jX_g + R_l + jX_l|} \tag{5.10}$$

Obviously, if the fault occurred nearer to the generator, the total impedance would be less and the fault current would be greater.

This equation gives I_{sc} as a scalar value, which is normal practice. For some purposes though, the X-on-R ratio of the impedance is also calculated.

5.7.3 Fault Currents in General

In a large power system of course, there are many synchronous generators and many transmission lines interconnecting them. The *prospective fault current*, at any point on the network, will be the sum of contributions from all the generators. And, this prospective fault current will be different at different points of the network.

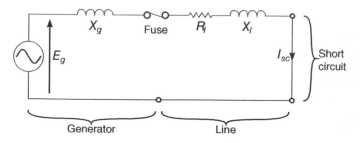

Figure 5.14 Single-phase equivalent circuit of a synchronous generator feeding a symmetrical short-circuit fault.

Additionally, in practice, many faults are not symmetrical and a full analysis requires calculation of the unbalanced currents, usually employing the method known as *symmetrical components*.

The calculation of fault currents is mathematically similar to load-flow and most power-system simulation software packages will include *fault analysis* as standard.

5.7.4 Fault Level (Short-Circuit Level) – Weak Grids

In systems operating at various voltages, it is useful to multiply the fault current I_{sc} by the nominal operating voltage (before the fault).

On a *per-phase* basis this is: $V_{ph}I_{sc}$, where V_{ph} is phase-to-neutral voltage.

In balanced three-phase system, $V_{ph} = \frac{V}{\sqrt{3}}$, where V is line-to-line voltage.

And multiplying by three, gives the three-phase *fault level*, S_k:

$$S_k = \sqrt{3}VI_{sc} \qquad (5.11)$$

where V is the nominal line-to-line voltage (before the fault) and I_{sc} is the symmetrical three-phase fault current as introduced earlier.

The fault level is a slightly abstract quantity since the voltage and current being multiplied together are not occurring at the same time. It does *not* indicate the amount of power available. Its units are VA, kVA or MVA, not kW or MW. The usefulness of the concept is that it gives the rating of a circuit breaker capable of withstanding the full voltage when the system is normal and of interrupting the highest prospective fault current at the point at which the fault level is calculated. As a consequence, the numerical value of the fault level and therefore the rating of circuit breakers is very large indeed. Table 5.2 gives typical values of fault levels at various system voltages.

The fault level at a particular point will vary as generators are brought on or taken off line elsewhere in the network. Also, it will vary if the network is reconfigured. Given that the entire protection system is designed around calculated fault levels, network operators must be aware of changes in fault levels and manage these appropriately.

The fault level is an important design parameter, not only for predicting currents under fault conditions, but also for predicting performance under normal operating conditions. The Thévenin equivalent circuit in the following section illustrates this concept (Figure 5.15).

Table 5.2 Typical fault levels

Nominal system voltage (kV)	Fault level (MVA)
132	5 000–25 000
33	500–2 500
11	10–250

Figure 5.15 Thévenin equivalent circuit.

5.7.5 Thévenin Equivalent Circuit

Consider again the Icelandic national power system shown in Figure 5.12. It is often useful to be able to calculate performance at one particular node without having to perform a full load-flow or fault analysis of the entire network. This can be achieved by representing the network by its *Thévenin equivalent* as explained in detail in Appendix A.16.

The values in the Thévenin equivalent circuit must be calculated for the particular node in question.

The *magnitude* of Z_{th} can be found from the fault level at the node by:

$$|Z_{th}| = \frac{V^2}{S_k} \tag{5.12}$$

where V is the nominal line-to-line voltage.

The *angle* of Z_{th} could also have been found from the fault level, if only I_{sc} and S_k had been calculated and stored as complex numbers. Instead of this, normal practice is to express the fault level as a scalar and to express the associated angle as an X/R ratio, which can be used in: $Z_{th} = R_{th} + jX_{th}$. The Thévenin source voltage can often be taken as the nominal voltage at the point of interest being careful to use the phase-to-neutral or line-to-line value consistent with the calculation.

The fault level is an important design parameter, not only for predicting currents under fault conditions, but also for predicting performance under normal operating conditions. It defines the *strength* of the network at a particular point. A *weak grid* is a network or part of a network where fault levels are low; that is, the Thévenin or source impedance Z_{th} is high. A high Z_{th} implies that the node voltage would be sensitive to active and/or reactive power extraction or injection at the node.

The impact of a renewable energy generator on a network is very dependent on the fault level at the point of connection. Considering a proposed wind farm capacity (in MW) as a percentage of the fault level (in MVA) can provide a rough guide to acceptability. The topic is dealt with in greater detail in Section 6.1.3.

5.8 Time Varying and Dynamic Simulations

One load-flow calculation (after it has iterated and converged) provides the results for one snapshot of power-system operating conditions. Many power-system simulation packages allow the user to define load and generation profiles: time-series data at daily, hourly or shorter intervals. The load-flow is then run repeatedly to provide voltage profiles, loading profiles and so on, on the same time-base.

Reducing the time-step to milliseconds allows *dynamic simulations* to be performed, in which the dynamic characteristics of the generators and their associated control systems are included. This type of dynamic simulation is used for stability studies, particularly in the design of transmission systems. The behaviour of the network is modelled through repeated load-flow calculations similar in principle to those discussed throughout this section. Load-flow uses phasor representations of voltages and current and therefore implicitly assumes pure sinusoidal operation.

More sophisticated simulation packages offer *transient analysis*, in which the components are represented by differential equations. This allows non-sinusoidal operation to be modelled but is very much more computationally demanding. Transient simulation is used to model lightning and other fault conditions requiring very detailed analysis of systems with relatively few nodes.

Many power-system simulation packages now include facilities to model renewable energy generators such as wind turbines and wind farms. A few years ago, the models tended to be little more than a 'negative load', which is adequate for basic load-flow calculation but not much else. Today, the leading packages include much more sophisticated models of wind turbines, intended for use in fault analysis and dynamic simulations. But, the dynamic behaviour of wind turbines, particularly those using doubly-fed induction generators (DFIGs), is very dependent on the detail of the generator control software. The turbine manufacturers are reluctant to provide these details and this has hampered the development of software simulation models. Some simulation software houses claim to have obtained sufficient details, but validation of these models remains an issue. Further details on dynamic modelling software and the representation of renewable energy sources can be found in Chapter 6.

5.9   Power-System Stability

One aspect of power-system stability has already been dealt with in Chapter 3, it was touched on in Section 4.2.8 in the context of synchronous machines and also in Section 5.8.

Power-system stability is an essential property of a power system that allows it to remain at or near a viable and acceptable equilibrium state during normal operation, and to return to that state if subjected to a disturbance due to some event such as the short circuit of a transmission line or the sudden loss of major plant. Traditionally, stability has concentrated on the connection of a synchronous machine to the power system as discussed in Section 4.2.8, but the concept is wider than this and involves all major elements in the power system that affect the system dynamics, including voltage regulators, transmission line characteristics, protection systems and so on. Three different forms of stability/instability can be identified: *rotor angle stability*, which is associated with dynamic variations in the load angle of synchronous machines as discussed in Section 4.2.8 and is sometimes referred to as first-swing stability (if pole slipping is to be avoided); *voltage stability*, which in practice is associated with meeting local dynamic demands for reactive power following a disturbance so as to maintain acceptable voltages throughout the power system and *frequency stability* that refers to the ability of a power system to restore frequency to an acceptable level following a severe mismatch between generation and load without disconnection of significant load or tripping of major generation (see Chapter 3). A complete analysis of stability can be found in Reference [4].

5.9.1 Equal Area Stability Criterion

Building on the analysis of the synchronous machine developed in Chapter 4, it is possible to derive a simple estimate of the maximum load angle following a transient mismatch of load, and thus assess the first-swing stability.

Recall from Chapter 3 that the inertia constant of a synchronous generator is defined as KE/P_{GMAX}. It can be shown [4] that in the case of no damping:

$$\frac{2H}{\omega_{NOM}} \frac{d^2\delta}{dt^2} = P_m - P_{max} \sin \delta \tag{5.13}$$

where P_m and P_{max} are the mechanical power input and the maximum electrical power output, respectively, both in power per unit, meaning that they have been divided by a suitable base value to give a dimensionless measure.

Using a plot of power against load angle (as in Figure 4.8) the result of a disturbance (in this case a sudden increase in mechanical power, can be investigated).

Let it be assumed that the mechanical power input increases abruptly from an initial equilibrium value P_{m1} to P_{m2} as shown in Figure 5.16. Because of the inertia the rotor angle cannot increase immediately to a new equilibrium point at b but the difference between the mechanical and electrical torques will accelerate it in that direction. By the time it reaches b there will be no accelerating torque but the rotor speed will be above the synchronous speed. As the rotor moves towards c, the rotor is decelerating since the electrical torque given by the P, δ curve is then greater than the mechanical power. By the time c is reached, the synchronous speed is reached but the rotor continues to decelerate. Without any damping, as in this simple model, the rotor angle will oscillate between points a and c indefinitely.

From Eq. (5.13) the rotor angle can be related to the accelerating power:

$$\frac{d^2\delta}{dt^2} = \frac{\omega_{NOM}}{2H}(P_m - P_e) \tag{5.14}$$

Figure 5.16 Power angle variations during transient.

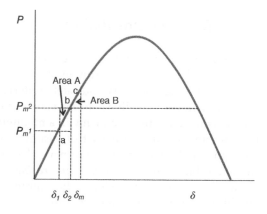

Since $\frac{d}{dt}\left[\frac{d\delta}{dt}\right]^2 = 2\frac{d\delta}{dt}\frac{d^2\delta}{dt^2}$, then multiplication of both sides of (5.14) and integration with respect to t results in $\left[\frac{d\delta}{dt}\right]^2 = \int \frac{\omega_{NOM}(P_m-P_e)}{H}d\delta$. For the system to be stable, the speed deviation, which is initially zero, must become zero again at some time after the disturbance when δ reaches its maximum value δ_m. At this point it follows that:

$$\int_{\delta_1}^{\delta_m} \frac{\omega_{NOM}(P_m - P_e)}{H}d\delta = 0 \tag{5.15}$$

This is true when areas A and B of Figure 5.16 are equal, hence the name of the stability criterion. The method can be used to assess whether there is a maximum value of δ for which this criterion applies.

The method is limited to the connection of a single generator to a power system and cannot deal with more complex oscillatory modes, such as inter-area modes, where one group of generators swings against another group, usually some distance away. Study of these more complex stability issues requires much more sophisticated models and these days commercial modelling packages are widely used for such purposes. Simplified representations of renewable energy sources are available within these packages and these will be discussed in Chapter 6.

5.9.2 Power-System Stabilisers

Power-system stabilisers make use of the excitation control of large synchronous generators to help damp power-system oscillations thereby improving the dynamics and increasing power-system stability. Rotor oscillations arise from the characteristics of synchronous generators as described in Section 5.9.1. These oscillations are too fast to be controlled by the prime mover governors. However, voltage regulators are fast enough to play this role. The machine excitation is adjusted to provide a torque that is in phase with rotor oscillations and this is achieved by using additional control signals to derive the excitation current; most importantly, the generator speed deviation. Of course, account must also be taken of the excitation dynamics, in particular the phase lag between the excitation and the generator torque. Full details of the design of power-system stabilisers can be found in Reference [4].

5.10 Dynamic Line Rating

In traditional power-system design and operation, overhead line conductors are assumed to have a fixed maximum allowable current, the static line rating (SLR), to meet given design, security and safety criteria. In practice this limit is determined by the allowable sag of the line between support points that occurs as a result of heating of the line, due to the need to maintain a minimum ground clearance. The SLR is derived by assuming a conservative set of weather conditions, such as high air temperature and light winds and through a thermal balance, calculating the conductor sag and thus the allowable current. *Dynamic line rating* (DLR) is a relatively new approach that allows the ampacity of the line to be adjusted in real time to reflect changing ambient conditions, in particular, temperature, solar radiation and wind speed and direction,

[5]. In this manner, the allowable current can often be increased above the SLR. This is particularly relevant to wind energy as wind cooling of the overhead line will be higher during periods of high wind, and a higher current can be allowed, permitting the connection locally of increased wind generation capacity [6]. Recent research has concentrated on forecasting of conditions and probabilistic assessment so as to allow the DLR to be based on a thorough risk assessment [7].

5.11 Reliability Analysis

The ability of an electrical power system to satisfy the demand with a reasonable probability of continuity is a measure of reliability. Loss of electricity supply may result in significant economic and social impact to consumers. Maintaining a reliable supply is therefore of paramount importance to electric generators and distributors. The subject of power-system reliability evaluation is one that has been studied extensively for traditional generating systems. The evaluation of reliability when distributed generators are also present complicates the issues and is a topic that is gaining importance. Distributed generators possess variability that may be beneficial, for example, their statistical output is correlated to local demand or the converse in which case they may contribute disbenefit under certain conditions. The general unpredictability that characterises most RE sources requires that past reliability assessment techniques based on deterministic principles must be modified to account for the probabilistic nature of embedded generation. This topic is beyond the terms of reference of this book but more information can be found in Reference [8].

References

1 National Grid Company (NGC) website http://www.nationalgrid.com/uk. In particular, the *Seven Year Statement* is very informative and now includes a large section on *Embedded and Renewable Generation*.
2 Gross, C.A. (1986). *Power System Analysis*. Wiley.
3 Wallis, E.A. (ed.) (1991). *Modern Power Station Practice, Volume L, System Operation*. Oxford: Pergamon Press.
4 Kundur, P. (2011). *Power System Stability and Control*. McGraw Hill.
5 Jupe, S.C.E., Bartlett, M., and Jackson, K. (2011). Dynamic thermal ratings: the state of the art. In: *21st Int. Conf. Electricity Distribution, CIRED*, no. 918,, 1–4. CIRED.
6 Fernandez, E., Albizu, I., Bedialauneta, M.T. et al. (2016). Review of dynamic line rating systems for wind power integration. *Renewable Sustainable Energy Rev.* 53: 80–92.
7 Fan, F., Bell, K., and Infield, D. (2017). Probabilistic real-time thermal rating forecasting for overhead lines by conditionally heteroscedastic auto-regressive models. *IEEE Trans. Power Delivery* 32 (4): 1881–1890.
8 Jenkins, N., Allan, R., Crossley, P. et al. (2000). *Embedded Generation*. Institution of Electrical Engineers.

6

Renewable Energy Generation in Power Systems

6.1 Distributed Generation

6.1.1 Introduction

Figure 3.2 shows that the maximum demand of England and Wales, a typical developed European country, is of the order of 50 GW. Most of this power is supplied by gas, coal and nuclear fuelled, combined cycle and steam driven generator sets rated up to 500 MW. The entire peak load, therefore, could be met by a hundred or so sets. As most power stations have two, three or four such generator sets, only a few tens of power stations are required to supply the many millions of consumers that are connected to the system. This type of generation is known as *centralised generation* and is typical of large power systems worldwide. Centralised generation is normally connected to the transmission system [1–9].

The term *distributed generation* or *embedded generation* refers to smaller generators that are usually connected to the distribution network. Distributed generation includes:

- generators powered from renewable energy (RE) sources (except large scale hydro and the largest wind farms);
- combined heat and power (CHP) systems, also known a co-generation (co-gen);
- standby generators operating grid-connected, particularly when centralised generation is inadequate or expensive.

The focus here is on renewables, but much of this chapter applies equally to all three types of distributed generation.

Smaller generators are not connected to the transmission system because of the cost of high voltage transformers and switchgear. Also, the transmission system is likely to be too far away since the geographical location of the generator is usually predetermined by the availability of the renewable energy resource. Such generators are *embedded* in the distribution network. An example of an embedded generator in a distribution network is shown in Figure 6.1.

Distributed generation alters the power flows in distribution networks and breaks the traditional *one-way power flow* from the high to the low voltage layers of the power system. In some circumstances, a distributed generator can be accommodated into a distribution network without difficulty, but occasionally it can cause a variety of problems. This chapter examines these issues.

Renewable Energy in Power Systems, Second Edition. David Infield and Leon Freris.
© 2020 John Wiley & Sons Ltd. Published 2020 by John Wiley & Sons Ltd.
Companion website: www.wiley.com/go/infield-renewable-energy

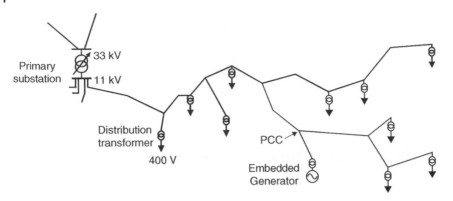

Figure 6.1 Example embedded generator.

6.1.2 Point of Common Coupling (PCC)

A basic requirement in connecting any generator to a power system is that it must not adversely affect the quality of electricity supplied to other customers on the network. With this in mind, it is useful to identify the *point of common coupling* (PCC).

Official definitions vary, but in simple terms the PCC is the point where the generator is connected to the public network as shown in Figure 6.1. In other words, the PCC is the point on the network, nearest to the generator, at which other customers are, or could be, connected. Thus, the exact location of the PCC can depend on who owns the line between the generator and the rest of the network. The importance of the PCC is that it is the point on the public network at which the generator will cause most disturbance.

6.1.3 Connection Voltage

The distribution system in the UK includes 400 V, 11 kV, 33 kV and 132 kV; other countries have similar voltage levels. As mentioned in Section 5.7.4, the fault level at the point of connection, which is a measure of network strength, is an important design parameter, not only for predicting currents under fault conditions, but also for predicting performance under normal operating conditions and, in particular, voltage rise.

The fault level at the PCC is very important when considering connecting a generator because it largely determines the effect that the generator will have on the network. A low fault level implies a high network source impedance, and a relatively large change in voltage at the PCC caused by extraction or injection of active or reactive power. The impact of a renewable energy generator on the network is therefore very dependent on the fault level at the point of connection as well as on the size of the proposed generator.

The appropriate voltage at which to connect a distributed generator is largely dependent on its rated capacity. There are many other factors, as will be seen, and so a range of indicative figures are used as guidelines. Of course, whether a network is weak or not is entirely in relation to the size of the generator being considered. It is therefore common to express a proposed renewable energy source capacity (in MW) as a percentage of the fault level (in MVA) that can be labelled 'short-circuit ratio'. This can provide a rough guide to acceptability. Typical figures for wind farms range from 2 to 24%.

Table 6.1 Design rules.

Network location	Maximum capacity of embedded generator
Out on 400 V network	50 kVA
At 400 V busbars	200–250 kVA
Out on 11 kV or 11.5 kV network	2–3 MVA
At 11 kV or 11.5 kV busbars	8 MVA
On 15 kV or 20 kV network and at busbars	6.5–10 MVA
On 63 to 90 kV network	10–40 MVA

Source: Taken from Jenkins, N., Allan, R., Crossley, P., Kirchen, D. and Strbac, G., Embedded Generation, IEEPower and Engineering Series 31, Institution of Electrical Engineers, London, 2000. Reproduced with permission of IET.

Table 6.1 gives rough figures for the maximum capacity of a generator that can be connected at a particular voltage level. These rules should ensure that the influence of the generator on the voltage at the point of connection is acceptable. Connecting at a higher voltage is usually more expensive because of the increased costs of transformers and switchgear and most likely a longer line required to make connection with the existing network. Connecting at too low a voltage may not be allowed if the generator were to result in an excessive effect on the local network. This can lead to a situation where the developer of the renewable energy system wishes to connect at one voltage level for economic reasons, while the network operator suggests connection at the next level up.

6.2 Voltage Effects

6.2.1 Steady State Voltage Rise

The connection of a distributed generator usually has the effect of raising the voltage at the PCC and this can lead to overvoltages for nearby customers. The need to limit this voltage rise, rather than exceeding the thermal capacity of the line, often determines the limiting size of generator that may be connected to a particular location.

An initial estimate of the voltage rise caused by connection of a generator can be obtained from analysis of the system as represented in a simplified form by Figure 6.2. The network up to the PCC can be represented by the Thévenin equivalent with the Thévenin impedance Z^{th} estimated from the fault level and X-on-R ratio at the PCC.

From Eq. (5.11) the magnitude of Zth is given by $|Z^{th}| = V^2/S^k$ where V is the nominal line-to-line voltage. The X-on-R ratio is then used to find the resistance and reactance from $Z^{th} = R + jX$.

The voltage rise ΔV can then be found using Eq. (5.6): $\Delta V \approx (PR + QX)/V$, where P and Q are positive if the active and reactive powers are positive; that is, have the directions shown in Figure 6.2. If an induction generator is connected directly to the network, Q has a negative value.

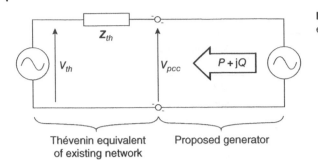

Figure 6.2 Equivalent circuit for estimating voltage rise.

Thévenin equivalent
of existing network   Proposed generator

The allowable voltage rise is dependent on how the network is currently operated; that is, how close the existing voltages get to the allowable maximum. Typically, a rise of 1% would be a concern to the network operator. Voltage rise is often the main consideration for wind farms, which tend to be in rural areas, connected by long and relatively high impedance lines. Voltage rise often puts a limit on the amount of generation that can be connected in a particular rural area. This can occur long before there is any chance of the power flow actually reversing or of the thermal limits of the lines being reached.

In some situations, calculations (or load flow modelling) show that voltages will exceed acceptable limits for only a few hours in a typical year. In this case, it may be cost effective to constrain generation during those hours. The lost revenue may be small in comparison to the cost of installing a stronger line.

Voltage rise can be mitigated through the extraction of reactive power at the PCC. With induction generators (the norm for smaller wind turbines), this can sometimes be achieved by removing some or all power-factor correction capacitors. With synchronous generators (more common in hydro and biomass fuelled systems), the excitation can be adjusted. However, an embedded generator will normally be charged for any reactive power that it consumes except in low voltage networks (400/230 V) where reactive power is not normally metered. Thus, the normal initial design objective is to operate with a power factor near to unity.

6.2.2 Automatic Voltage Control – Tap Changers

The simple picture presented so far is complicated in practice by the existence of automatic voltage-control mechanisms in distribution networks. The most common mechanisms are automatic on-load tap changers, which are fitted to most transformers throughout the system, except the final distribution transformers (in the UK, the 11 kV to 400/230 V transformers are as described in Section 5.5.1).

In the network shown in Figure 6.1, the primary substation is equipped with an automatic voltage controller, which uses a tap changer to adjust the turns-ratio of the transformer. Closed-loop control ensures that the voltage on the transformer secondary is kept close to 11 kV. This compensates for variations on the 33 kV side and the current-dependent voltage drops within the transformer itself.

Automatic tap changers affect the steady state ΔV voltage-rise analysis presented in the previous section. In particular, the Thévenin impedance Z^{th} must be adjusted, because its calculation was based on the fault level, which naturally included the impedance upstream of the voltage controller. With the automatic voltage controller active, the Thévenin voltage is the fixed voltage at the transformer secondary. To correct for this, the fault level and X-on-R ratio at the fixed voltage node (transformer secondary) must be known so that the upstream impedance may be calculated and deducted appropriately.

To complicate matters further some automatic voltage controllers do not simply keep the voltage constant. For example, some provide line-drop compensation, which is a way of estimating and allowing for the voltage drop in the downstream line. Other controllers use a technique known as negative reactance compounding, which allows dissimilar transformers or transformers at separate substations to be operated in parallel. Unfortunately, controllers using these techniques can be affected by the changes in the substation power factor that can be caused by distributed generation. To prevent such changes in the power factor, some embedded generators are designed to operate at the same power-factor as the typical loads. Problems of this nature could be avoided if the distributed generators are of a type that could generate and control reactive power. In practice, this option has not been fully exploited in the low voltage distribution networks, but with the ongoing development of more sophisticated and low-cost power electronic interfaces it is likely to be of importance in the future.

Since wind power is now predominantly connected through power electronics, and all PV generation is, this capability exists. Often converters operate at unity power factor, particularly with PV, and thus are passive in terms of local voltage control, but this need not be the case.

6.2.3 Active and Reactive Power from Renewable Energy Generators

As mentioned earlier, the large generators of conventional power systems have their individual automatic voltage regulators (AVRs) set to maintain the generator bus voltage virtually constant.

Generators fed from renewable energy sources are substantially smaller in rating and are often connected to the distribution rather than the transmission network. For these reasons, conventional generator control schemes are not always considered appropriate for small embedded renewable energy supplied generators. For example, a fixed speed wind turbine driving an induction generator is expected to inject into the network whatever power is converted from the wind up to the rated wind speed, beyond which the power is limited by aerodynamic means at the rated output. It is also expected to absorb whatever reactive power the induction generator requires from the network, minus any locally generated Q from power factor correction capacitors. It should be noted that such direct connected wind generators are no longer common. The situation, however, with small hydro systems is that these are still often interfaced to the grid through induction generators.

Energy from photovoltaic, wave and tidal schemes and MW size wind turbines will invariably be fed to the grid through a pulse-width-modulation (PWM) power electronic converter. This provides the facility of reactive power injection/extraction at the point of connection.

To summarise, for relatively small embedded renewable energy generators connected to strong networks the P injection depends solely on the renewable energy source (wind, sun, water) level at the time and the Q injection either on the natural generator Q/P characteristics or on the control characteristics of the power electronic converter. In the latter case, the converter could be regulated to inject active power at unity power factor, thus avoiding any Q exchanges with the network, and charges for reactive power.

It is also possible that a mutually beneficial formula could be agreed between the owner of the renewable generator and the utility so that the Q generated/consumed by the converter is adjusted to suit the local network.

Generators that are very small individually have negligible influence on system frequency. However, when large numbers of such generators are connected their aggregate impact can be significant. In fact, as penetration levels by wind power increase, as for example in Denmark where the annual level exceeds 40%, and much higher levels will occur at times, utilities will require that renewable generators are designed to contribute positively to power system frequency and voltage stability during contingencies, but more of this later. In a similar manner, it might be expected that in future scenarios with substantial PV generation, that the PV converters might be required to be controlled in relation to system frequency.

6.2.4 Example Load Flow

As discussed already, renewable energy generation can affect both line loadings and voltages throughout the system. Load flow is a technique that allows the flows of real and reactive current throughout the network to be calculated, based on the location of the loads and sources and the line impedances.

The network of Figure 6.1 is used here to illustrate the way in which load flow analysis is applied at a distribution level to assess the effect of connecting a renewable energy generator at a node. This network is redrawn in Figure 6.3 to include node numbering and a proposed embedded renewable energy generator at node 10. It is necessary to know whether such an embedded generator is likely to affect adversely the network voltage profile.

At the outset a fault analysis is undertaken to provide the short-circuit levels at all nodes and so indicate the acceptable rating of the proposed renewable energy generator. To embark on a load flow, it is necessary to specify the parameters of the lines, transformers and the known node variables. Table 6.2 gives the line data in terms of line position, length and line type. Table 6.3 provides additional information for each line type in terms of impedance and current rating.

Section 5.6.5 described the analysis required to determine network fault levels. For the network of Figure 6.3 the results are shown in Figure 6.4. As expected, the fault level is highest close to the secondary of the distribution transformer (node 1) and due to the

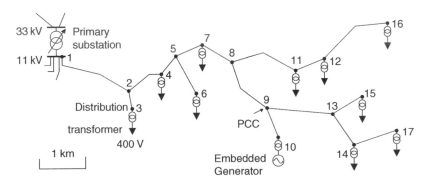

Figure 6.3 Example load flow.

Table 6.2 Line data (input to load flow).

From node	To node	Length (m)	Line type
1	2	1297	100AAAC
2	3	304	50AAAC
2	4	626	100AAAC
4	5	391	100AAAC
5	6	738	185AL
5	7	492	100AAAC
7	8	583	100AAAC
8	11	1110	50AAAC
11	12	539	50AAAC
12	16	1253	185AL
8	9	1000	50AAAC
9	10	539	185AL
9	13	1154	50AAAC
13	15	583	50AAAC
13	14	652	50AAAC
14	17	791	95AL

AAAC, All aluminium alloy conductor.

Table 6.3 Line type data (input to load flow).

Line type	R (Ω/km)	X (Ω/km)	Rating (A)
50AAAC overhead line	0.550	0.372	219
100AAAC overhead line	0.277	0.351	345
95AL underground cable	0.320	0.087	170
185AL underground cable	0.164	0.085	255

cumulative impedance of the transmission lines declines moving away from this point. At node 10, where the renewable energy generator is to be connected, the fault level is 42 MVA. A wind turbine rated at 500 kW resulting in a short-circuit ratio of $0.5/42 = 12\%$ is within the acceptable range.

In this example, it is assumed that the embedded generator is a wind turbine consisting of a directly connected induction generator operating at a power factor of 0.9 at full output. At a rated output this provides an injection of 500 kW and the absorption of 250 kVAR at node 10. In order to determine the voltage profile of the network under these conditions there is a need to assess whether this proposal is feasible. To proceed with the load flow analysis it is required that two variables are defined at each node. This distribution network consists solely of consumer PQ nodes and one embedded generator that can also be treated as a PQ node. The input data to the load flow are shown in Table 6.4.

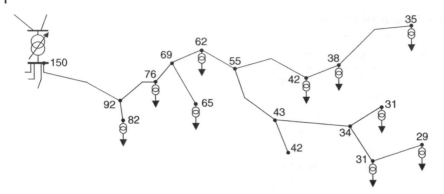

Figure 6.4 Example fault levels (in MVA) throughout a typical rural 11 kV feeder.

Table 6.4 Node data (input to load flow).

Node number	P (kW)	Q (kVAR)
1		
2		
3	−238	−71
4	−159	−48
5		
6	−340	−102
7	−178	−53
8		
9		
10	500	−250
11	−458	−137
12	−221	−66
13		
14	−97	−29
15	−386	−116
16	−161	−48
17	−64	−19

The results of the load flow analysis are shown in Figure 6.5. The voltage profile at consumer nodes is easily within the acceptable limits of ±1%. At the PCC, the rise of voltage due to the active power injection is in this case moderated by the extraction of reactive power from the renewable energy source. Of course, if capacitors are installed at the wind turbine to correct the power factor towards unity, the voltage rise would have been significantly higher and perhaps unacceptable. A rerun of the load flow analysis would provide a wealth of information on system voltage sensitivity to a range of renewable energy source locations and operational conditions.

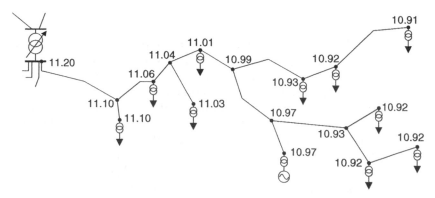

Figure 6.5 Load flow results: voltages (kV).

6.3 Thermal Limits

Power injected by a distributed generator will alter the local network power flows. Power flows in individual components may be increased, decreased or even reversed. Power flow (both active and reactive) requires current flows and this causes heating in lines, cables and transformers. Each of these components has a thermal limit, which effectively determines the maximum current it can carry, and these thermal limits can set an upper limit on the size of the renewable energy generator that can be accommodated. Load flow analysis allows all the component currents to be calculated.

6.3.1 Overhead Lines and Cables

It may appear surprising that a great many lines and cables are normally operated at well below their thermal limits, particularly in lower voltage networks (up to say 20 kV). There are two reasons for this. First, in order to keep customer voltages within range, voltage drops in conductors must be limited, which often requires use of larger conductors than those dictated by thermal limits. Second, energy losses are significantly less in larger conductors and a life-cycle cost analysis usually favours such oversizing. As a consequence thermal limits of lines and cables are rarely reported as a limiting factor in the installation of embedded generators in 11 kV or similar networks. Voltage and increased fault level considerations arising from the connection of additional generation, discussed later, usually come into play first.

At higher voltage levels, that is, 33 and 132 kV, voltage control is less demanding and thermal limits of lines and cables may well be the limiting factor. These limits are well understood and rarely the subject of dispute, but there are a couple of aspects worth noting. First, care is needed regarding the demand that is assumed to be concurrent with the generation in calculations. Ideally, this should represent the worst-case scenario and does not necessarily occur when the maximum expected input from the renewable energy source(s) is concurrent with the maximum demand. Second, overhead lines are naturally cooled by wind, which increases their capacity during periods of high wind. Consequently, lines installed primarily to carry wind power can be undersized relative to conventional design guides (dynamic line rating was discussed in Chapter 5).

6.3.2 Transformers

Determining the thermal limits for transformers is more complicated. The rating of a transformer is generally given in kVA or MVA, but there may be three different figures quoted, corresponding to natural (convective) cooling, fan cooling, and cooling by pumped oil. Furthermore, transformers have long thermal time constants and thus can be overloaded for short periods of time without causing overheating or significant damage. Calculation of the damaging effect of such transient temperatures can be difficult.

Transformers are generally chosen to match the expected maximum demand and are normally operated at a significant proportion of their thermal limit, partly because standing (no load) losses are significant and so the efficiency of a lightly loaded transformer is poor. Thus, in areas with very high penetrations of embedded generation, thermal limits of transformers can restrict further installation. In round figures, this only occurs when the rating of an embedded generator connected below a particular transformer, less the minimum load in the same area, exceeds the maximum load in that area.

Most transformers can accommodate reverse-power flow, up to the normal forward rating, without a problem. However, there are some (not many) on-load tap changers that have very limited reverse-power capability. Furthermore, some AVRs (control relays) associated with on-load tap changers can be affected by reverse-power flow.

Thus, the thermal limits of existing transformers are not normally a limiting factor in the installation of distributed generators, unless the installed capacity of generators exceeds the maximum demand in the area served. Such high penetrations are not common, at least in the UK, although there is one example in Wales where two wind farms are connected below one substation. Power flow through this substation is frequently reversed and occasionally generation has to be curtailed when the reverse-power flow reaches the transformers' thermal limits.

6.4 Other Embedded Generation Issues

Voltage flicker, steps, dips, harmonics and phase imbalance may all degrade what is loosely described as *power quality*. Such undesirable occurrences can be caused just as much by loads as by generators. In general, it is the responsibility of the operator of the load or generator not to unduly affect the power quality for other users in the area. However, the impact of any load or generator is dependent not only on its own characteristics but also on the strength of the network to which it is connected.

Network operators are paying increasing attention to power quality. This is not specifically in relation to distributed generation but mainly because of regulatory obligations set up to protect increasingly discerning consumers from the generation of harmonics caused by the plethora of electronic devices on the network.

6.4.1 Flicker, Voltage Steps and Dips

6.4.1.1 Flicker
Incandescent light bulbs can cause annoyance by flickering due to rapid voltage variations. The human eye is particularly sensitive to flicker at around 8 Hz. Sensitive

electronic equipment may also be affected. Such voltage variations are generally caused by rapid variations of active or reactive power flows within the network. The classic causes of unacceptable flicker are sawmills in rural areas and arc furnaces. Distributed generators can also contribute to flicker. A flicker meter measures voltage variations of this type and gives figures for short and long term severity.

Wind turbines can contribute to flicker because tower shadow and turbulent winds can cause rapid variations in active and reactive power output. Fixed speed pitch-regulated turbines are generally worse than stall-regulated ones at high wind speeds. This is because the pitch regulation action in response to, say, a gust is slow to take effect due to limited pitch actuator capability and the inertial mass of the blade, while the aerodynamic action of stalling a blade is very fast indeed. However, modern multi-megawatt wind turbines are designed to operate at variable speeds up to the rated wind speed, and also to allow marginal speed variations between rated and cut-out wind speeds. This results in smoothing of the power delivered, especially around rated power, and a substantial reduction in flicker.

The degree of flicker caused by a given generator is highly dependent on the network characteristics, in particular the fault level and X/R ratio at the PCC. Thus, a generator that is perfectly acceptable at one point on a network may be unacceptable elsewhere. In general, this dependence is the same as that for a steady state voltage rise; thus, a wind farm that causes a significant voltage rise may also cause significant flicker. Flicker is more of an issue for single turbines or small clusters on low voltage distribution networks. As the power fluctuations from individual wind turbines are not correlated, the effect of flicker is less pronounced for large wind farms.

Incandescent light bulbs are much less frequently used, indeed they can no longer be sold in European Union countries because of their inefficiency. Other forms of lighting are much less sensitive to voltage variations but despite this, Standards still exist to limit flicker (for example, the International Standard IEC 61000-3-3).

6.4.1.2 Steps and Dips

Switching a generator, either on or off, can cause a step change in the voltage at the PCC due to an associated abrupt change in active or reactive power flow. If the power changes associated with switching can be made gentler, then such voltage steps can be minimised, at least during planned switching operations.

Direct on-line starting of an induction generator can cause a dip in the voltage, due to large starting currents sometimes called inrush currents. For large induction generators, it is normal to use a *soft-starter* to limit these starting currents. Both steps and dips, if repetitive or cyclic, may contribute to flicker. With the increasing use of power electronic interfaces between the renewable energy generator and the grid, steps and dips during switching are virtually eliminated.

6.4.2 Harmonics/Distortion

Both loads and generators can cause distortion of network voltages, and such distortion is usually expressed in terms of harmonics. A common source of harmonics is the widespread use of cheap power supplies found in computers, televisions and other appliances. The voltage waveform in a modern office building or in a residential distribution network can be badly distorted by harmonics created by such loads, long before any distributed generators are even contemplated.

The distortion of the voltage results from nonsinusoidal currents flowing through the effective network impedance, which is related to the Thévenin impedance introduced earlier. The Thévenin impedance as calculated in Section 5.7.5 cannot be used directly to analyse harmonics because it is only accurate at 50 Hz. Experiment and analysis can provide a curve relating the system Thévenin complex impedance as a function of frequency. Such curves can be used in conjunction with the known levels of generated harmonics from say, a converter, to predict the likely distortion of the PCC voltage by a renewable energy embedded generator. Voltage distortion will usually be worse if the PCC node fault level is low.

In theory, a poorly designed directly connected synchronous or induction generator could cause significant harmonics, but modern machines manufactured by reputable companies are generally free of such deficiencies. Harmonics from distributed generators these days usually arise from their power electronics. Some early wind turbines and PV systems used thyristor based converters and did cause significant harmonics, but this is no longer the case.

Modern PWM converters as applied to grid-connect PV and with variable speed wind turbines (using Metal Oxide Semiconductor Field Effect Transistors (MOSFETs) and Insulated Gate Bipolar Transistors (IGBTs)) are very much better in this respect but still have to be carefully designed and checked for suitability in each installation. In particular, there is a trade off because a high switching frequency will reduce harmonics but will increase switching losses and therefore reduce the converter efficiency.

A particular Spanish region with a very high penetration of wind power was found to have excessive harmonic distortion due to PWM converters with a rather low switching frequency. There is concern that very high penetrations of PV could have a similar effect, though this should be considered in the context of the rapid increase of power electronics in consumer appliances. National and International Standards, for example EN 61000-3-2: A, specify strict limits for the harmonic distortion allowed from small loads and generators, and network power quality in this regard is not regarded as a particular concern.

6.4.3 Phase Voltage Imbalance

Ideally, single-phase loads are evenly distributed between the three phases. In practice, any differences will cause a phase voltage imbalance on the network.

A directly connected three-phase generator will normally serve to reduce any existing phase voltage imbalance. However, in achieving this, the generator itself will heat up due to abnormal circulating currents within the windings. To avoid overheating, wind turbines usually include *imbalance protection*. In some rural networks, this imbalance protection can operate frequently and cause significant loss of generation.

In small distributed applications of RE, for example domestic wind turbines or photovoltaics, single-phase generators could increase phase voltage imbalance. However, if such applications were to be adopted in large numbers, statistically they would be evenly distributed on average among the three phases just as loads are and no problems should occur.

6.4.4   Network Reinforcement

In some situations, the voltage effects discussed previously could be reduced by reinforcing the network and thereby increasing the fault level at the PCC. For example, it is sometimes possible to upgrade the conductors of an existing overhead line, which is known as reconductoring. Alternatively, if permissions can be obtained, it is usually more cost effective simply to build a new line.

When faced with an application to connect an embedded generator, it is common for the local electricity company to claim that the nearest point for connection to the existing network is too weak and to require that a new line be installed in order to reach a stronger part of the network. The cost of installing such a line can be very significant in comparison to the size and cost of the generator itself and will normally fall to the owner of the generator, although rules for such network reinforcement vary from country to country.

6.4.5   Network Losses

Enthusiastic proponents of renewable energy sometimes claim that, because distributed generation is nearer to the loads, it will reduce network losses. Unfortunately, this is not always true.

High-voltage transmission networks are generally very efficient; thus reducing power flows in transmission networks does not bring very large reductions in losses. In distribution networks, losses are proportionally far greater and so, for losses to be reduced to any significant extent, the distributed generation has to be located very close to matching loads. Furthermore, its availability profile in time should reasonably match that of the local demand so that power flows are minimised.

6.4.6   Fault Level Increase

As noted previously, fault level can be used as a measure of grid strength and can be compared with proposed generator capacity to give an indication of likely effects on-line voltage. However, as shown in Figure 6.6, the fault level can itself be increased by the connection of embedded generators, particularly where synchronous machines are involved. This does not invalidate the previous comments on voltage effects, but it does represent another factor to be considered in the connection of a generator.

Switchgear is sized to suit the network fault level. Existing switchgear therefore represents a limit to the amount of embedded generation that can be accommodated in a given area. It is not just the switchgear on the site of the embedded generator that is important, as a synchronous generator will increase the fault level throughout the local network and so the rating of all local switchgear must be considered.

In some cases, a proposed generator will make a relatively small contribution to the fault level but this could be enough to increase the existing fault level above the maximum allowable for a particular switchgear component. This component may be serving many other customers and be large relative to the generator. Upgrading this component could easily cost far in excess of the generator itself. A fault level increase is often the limiting factor for installation of embedded generation in urban areas where the lines are short and impedances low.

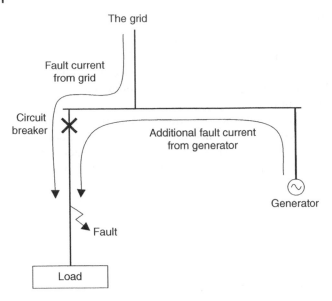

Figure 6.6 Increased fault current through a circuit breaker.

The good news is that the large majority of distributed renewable energy sources are not interfaced to the network through a directly connected synchronous generator but through a power electronic converter of the PWM type. Such converters consist of very fast switching elements and generally contribute negligible overcurrents during faults. In most cases there will be no need to upgrade the local circuit breakers.

Conversely, care must be taken to ensure that the inability of power electronic converters to provide fault current does not conflict with a protection system design that might rely for example on high currents to blow fuses.

6.5 Islanding

6.5.1 Introduction

A generator is said to be *islanded* if it continues to supply a local load after being disconnected from the main network. On a larger scale, an island can include many generators and many loads.

If islanding is to be permitted, then the system must be designed to be capable of operating both grid-connected and standalone. Generally speaking, the standalone mode is the more difficult to engineer since it must include generation/load matching in order to ensure that voltage and frequency are kept between acceptable limits. Generators that are designed to operate both grid-connected and standalone are often called *standby* generators and are installed primarily to provide security of supply to particularly important loads, such as hospitals and computer centres. Standby generators are usually powered from diesel or gas.

Systems employing renewables are rarely designed to operate both grid-connected and standalone. The extra cost of properly engineering a standalone mode is unlikely to be justified unless grid power cuts are very frequent.

If a generator is not designed to operate in a standalone mode, it must be capable of shutting down immediately when the grid connection is lost, otherwise the system may become dangerous for the following reasons:

- The island may not be properly earthed and may present an electric shock hazard to personnel.
- A system that appears to be disconnected from the mains but is in fact still live also presents a shock hazard.
- The voltage and/or frequency may deviate from the normal range and this may damage other equipment connected within the island.
- The island's frequency will quickly deviate from that of the main network and if the two networks are reconnected under these circumstances, equipment (particularly synchronous machines) will probably be damaged.
- The fault level within the island will be reduced and, if a fault were to develop, the resulting current might be insufficient to operate protective devices such as circuit breakers and fuses decisively.

6.5.2   Loss-of-Mains Protection for Rotating Machines

Satisfactory *loss-of-mains* detection is surprisingly difficult to achieve with rotating machines. The detection must be reliable, fast and stable and not prone to wrong tripping. The requirement that it must be fast stems from the widespread use of *auto-reclose circuit breakers* in distribution systems and from the fact that these are not normally fitted with synchronisation checking. Auto-reclosers are designed to restore a supply very quickly after a transient fault. Most faults in distribution systems (upwards of 80%) are transient and thus reclosers play a very important role in reducing customer-minutes lost. Reclose times in distribution systems can be as short as one second, which is good for consumers but also implies that loss-of-mains protection must be designed to operate well within one second.

For small induction generators, under/overvoltage and under/overfrequency protection is considered adequate protection against loss-of-mains (in the UK). For synchronous generators and large induction generators the preferred loss-of-mains detection techniques are:

- rate of change of frequency (ROCOF) and
- vector shift.
- Both techniques rely on there being a significant change in power flow within the island when the mains is lost. This will naturally cause (i) the generator to accelerate or decelerate (detected by ROCOF) and (ii) a change in the generator power factor (detected by vector shift). Unfortunately, there is no guarantee that there will be a significant change in power flow and so neither technique proves to be perfect in practice.

6.5.3   Loss-of-Mains Protection for Inverters

An inverter that is intended for grid connection will not normally be designed to island. It will be designed to convert all the power available at a given time from the generating source (PV for example). It will not be designed to supply a constant AC voltage to a variable load. Islanding is only possible if the load demand exactly matches the source supply, a very unlikely scenario. Nonetheless, the designs of such inverters usually include explicit loss-of-mains protection, indeed connection standards such at the UK's Engineering Recommendation G83/1 [2] can require this.

6.6 Fault Ride-Through

Prior to about 2001, all wind turbines were required to shut down immediately on the occurrence of any disturbance on the grid. This was primarily to reduce any risk of islanding, which can be dangerous. It was a sensible requirement in systems with very low wind power penetrations. Wind turbines were fitted with loss-of-mains protection relays triggered by any of undervoltage, overvoltage, underfrequency, overfrequency, ROCOF and vector shift incidents.

If such a protection relay is set to trip reliably on a genuine loss-of-mains, then it will also be prone to nuisance tripping when there is merely a transient disturbance on the mains. This raises the possibility of *block tripping* where a widespread transient disturbance across the network could cause a large number of loss-of-mains relays to trip simultaneously. Block tripping is highly undesirable in the context of controlling the voltage and frequency of the overall power system. The concern is that, following a disturbance on the grid, such as the tripping of a large generator, a substantial block of wind power would also trip off, which would aggravate the shortage of generation, making matters considerably worse for frequency control. Because of this, regulations are in place, particularly for very large wind farms that are to be connected to the transmission network. These regulations require that wind turbines do not trip unnecessarily, but rather that they will ride through transient disturbances and to stay connected and resume generation as soon at the fault is cleared.

The first requirement is for wind turbines not to be tripped off the grid during local faults that result in PCC voltage reductions. Wind turbines fitted with controllers capable of such behaviour have been tested and shown to be capable of riding through a fault and re-establishing full power immediately after the end of the fault. It is expected that increasingly sophisticated control functions are likely to be demanded from wind and other renewable energy generators as penetrations increase. Even with small but numerous embedded renewable generators in the future, it may eventually become important to coordinate their response both to steady state and transient network conditions. At present no such arrangements are in place.

In 2001, the German transmission system operator, E.ON Netz, issued new requirements including fault ride-through. Since then, most of the other major European transmission system operators have issued similar requirements, within their respective *grid codes*. The details vary between countries but the general requirement is as follows.

During a short circuit on the network near to the wind farm, the voltage at the connection point of the wind farm is assumed to follow the profile illustrated in Figure 6.7. The wind turbines must remain connected to the network; they should generate reactive power during the fault and should resume generation of active power as soon at

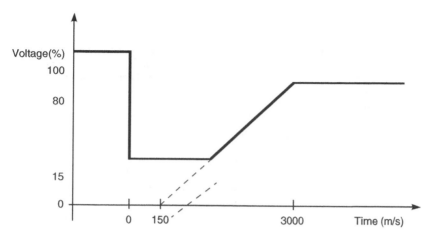

Figure 6.7 Fault ride-through voltage profile.

the fault is cleared. Meeting this requirement was, and still is, a major challenge for manufacturers of wind turbines, especially with certain generator configurations.

Wind turbines with full converters are readily redesigned to meet the fault ride-through requirement. Turbines with directly connected induction generators are inherently unsuited. The drop in terminal voltage causes the flux to collapse, the electromechanical torque almost disappears and the wind turbine accelerates rapidly. To prevent over speed, the brake has to be applied and the turbine cannot immediately resume generation when the fault is cleared. To address this (and for other reasons), Siemens added full converters to some of their models.

Doubly fed induction generator (DFIG) manufacturers also faced a considerable challenge. In order to protect the IGBT converters from excessive current during a fault, DFIGs are generally fitted with a crowbar, which is a set of thyristors that short out the rotor and cause the DFIG to behave like a squirrel-cage machine. All major manufacturers of DFIGs now claim to have developed their controllers such that they can ride through faults. In a competitive market, they do not publish the details, but it is understood that they generally employ an active crowbar, which provides a controlled short circuit to the rotor and allows the IGBT converters to continue operation.

The details of fault ride-through requirements differ from one country to the next, and there is concern that these details may be somewhat arbitrary. Meeting fault ride-through requirements adds costs to wind turbines and the industry is now concerned that these requirements should not be unnecessarily applied.

6.7 Generator and Converter Characteristics

Table 6.5 provides a summarised overview of the characteristics of synchronous and asynchronous generators and power electronic converters relevant to the issues discussed in this chapter. Further information on the topics of this chapter can be found in References [3–9].

Table 6.5 System-relevant characteristics of generators and converters.

	Rotating machines		Power–electronic converters	
	Synchronous	Induction (asynchronous)	Line-commutated (old – thyristors)	Self-commutated (modern – PWM)
Operating speed (rev/min)	Exactly synchronous speed	Above synchronous speed when generating, above and below synchronous in DFIG systems		
Reactive power	Either generate or consume; often controlled by an AVR, which may be set to give constant bus voltage or constant power factor	Machine *itself* will consume; may have power-factor correction capacitors fitted or may consume or generate if fitted with PWM converter	Consume	Either generate or consume; often designed to operate at unity power factor
Starting currents	Should be insignificant	Significant unless a soft-starter is used	None	None
Fault contribution	Significant	Usually insignificant	Insignificant	Insignificant
Synchronisation procedure	Yes	No	No	No
Risk of islanding	High (if completely unprotected)	Low but slightly increased when power-factor correction capacitors are fitted	None	Low (unless designed to)

References

1 National Grid Company (NGC) website http://www.nationalgrid.com/uk. In particular, the 'Seven Year Statement' is very informative and now includes a large section on 'Embedded and Renewable Generation'.

2 ER G83/1 is available from the Energy Network Association; www.energynetworks.org.uk.

3 McGrow, L. (2006). Beat the system. *IET Power Eng.* 20: 28–30.

4 Jenkins, N., Allan, R., Crossley, P. et al. (2000). *Embedded Generation*, IEE Power and Engineering Series 31. London: Institution of Electrical Engineers.

5 (2006). *Decentralising Power: An Energy Revolution for the 21st Century*. Greenpeace.

6 Povlesen, A.F. (2003). Distributed power using PV. *Renewable Energy World* 6: 63–73.

7 Godoy Simoes, M. and Farret, F.A. (2004). *Renewable Energy Systems*. Boca Rabon, FL: CRS Press.

8 Masters, G.M. (2004). *Renewable and Efficient Electric Power Systems*. Chichester: Wiley.

9 Fox, B., Flynn, D., and Bryans, L. (2014). *Wind Power Integration – Connection and System Operational Aspects*, 2e. IET.

7

Power System Economics and the Electricity Market

7.1 Introduction

This chapter reviews the economics of renewable power generation and the electricity market mechanisms that increasingly determine what generators are used, and when. A distinction is made between the capital and running costs and these are compared for renewable and conventional forms of generation. A major attraction of renewable energy (RE) sources is their limited environmental impact and in particular their low life cycle CO_2 emissions. External costs, that is, costs associated with an activity that are not borne by the party irresponsible (discussed in detail in Section 7.4), for renewable power generation will thus be lower than conventional generation, in particular fossil fuelled plant. It is therefore important to consider how the economics of power generation change if external costs are taken into account. In addition, and for small-scale embedded renewable energy generation, there are further benefits that should be taken into account.

A description is given in this chapter of how renewable energy generation is financially accommodated within an electricity system in which power from generation plant is dispatched to meet the varying electricity demand. The UK situation is reviewed to illustrate how renewable energy generation fits within a system that has evolved from being wholly state-owned and controlled to one operating within a privatised market with open competition and trading since this is a development that has affected many, if not most, power systems around the world.

Finally, the range of support mechanisms that are used by different countries to encourage the growth of renewable energy generation is briefly reviewed, alongside the refinement of the power system market, and in particular the extension of this to include ancillary services.

7.2 The Costs of Electricity Generation

7.2.1 Capital and Running Costs of Renewable and Conventional Generation Plant

In order to assess the economics of renewable energy generation, it is first necessary to assess the costs of building and operating such plant and to compare these with the costs of conventional generation. Installed costs and performance levels of renewable

Renewable Energy in Power Systems, Second Edition. David Infield and Leon Freris.
© 2020 John Wiley & Sons Ltd. Published 2020 by John Wiley & Sons Ltd.
Companion website: www.wiley.com/go/infield-renewable-energy

energy plant are broadly similar around the world since the specific technologies are generally only installed where an adequate resource exists, although of course resource variation means that no single figure can quantify the exact costs and performance of any technology. In the case of wind and solar energy, for example, the energy productivity (kWh per kW) of plant varies significantly with the resource. Solar installations near to the equator are generally more productive than those in, say, Sweden or New Zealand. Wind energy variations are more extreme; the windiest regions are New Zealand, the British Isles and Antarctica, while equatorial regions tend to have very low average wind speeds. Not surprisingly, solar installations are more common in areas with good solar radiation, and wind turbines are not widely deployed in the equatorial regions.

Table 7.1 summarises the key parameters associated with renewable energy technologies and conventional plant to facilitate comparison and also gives an indication of their range. These figures loosely reflect UK costs for fossil fuels and nuclear fuel as can be found on government web sites, but these should be similar in other countries due to these fuels being internationally traded. *Fuel cost* is zero for most types of renewable energy but not for energy crops where it is positive, nor for the waste-burning technologies where it can be negative if landfill costs are avoided. It must be emphasised that the costs quoted are indicative and certainly do not apply to all situations.

Table 7.1 indicates the three components of electricity generation costs:

- The capital component includes the cost of the plant, land acquisition (unless a rent is paid, in which case this is a running cost); grid connection, although in some European states, the utility has to bear this cost; and initial financing costs (as

Table 7.1 Indicative cost and performance data for renewable energy and conventional plant.

Technology	Capital cost (€/kW)	O&M costs (€ cents/kWh)	Capacity factor	Fuel cost (€ cents/kWh)
Onshore wind	1000–1500	1.0–3.0	20–40%[a]	—
Offshore wind	1500–2200	2.0–5.0	30–50%[a]	—
Biomass (energy crops)	1500–4000[b]	1.0–2.0	70–85%[c]	0.5–1.5
Landfill gas	450–1300	1.5–3.0	70–90%	—
Municipal waste	2300–6000	4.5–7.5[d]	70–85%	See note[e]
Geothermal	1500–4000	1.0–2.5	75–85%	
Photovoltaics	1000–3000	0.2–1.0	10–30%	
Hydro-power	1000–4000	0.2–0.5	30–90%	
Solar thermal power	4500–10 000	3.0–4.0	20–70[f]	
Gas (CCGT)	450–700	0.3–0.8	85%	5.2–6.8[g]
Coal	1000–1200	0.7–1.5	75–80%	3.0–3.5
Nuclear	2000–4000	0.5–1.2	85–90%	0.6–0.9

a) Depends strongly on wind speed.
b) High cost plant is more efficient.
c) Production falls after ca. 15 years.
d) Depends on mix of waste.
e) Cost negative (€~4.5–7.5 cents/kWh), but steady supply needed.
f) Depends on whether the plant includes integrated thermal storage.
g) Gas prices are particularly volatile.

opposed to repayment costs). The way that capital costs translate into a component of generation cost is discussed in the next section.

- The operating and maintenance (O&M) component includes insurance, rent, and rates set by the local administrative authority, as well as the costs of labour and materials used for operations and maintenance.
- Fuel costs.

7.2.2 Total Generation Costs

Capital costs alone are not a reliable guide to competitiveness. For example, the low capital costs of combined cycle gas turbine (CCGT) plant do not result in low electricity generation costs unless the cost of fuel is also low. Similarly, although geothermal plant has a high capital cost it also has a high load factor and may therefore have similar generation costs to gas or wind.

It is conventional to compare different technologies in terms of the levelised cost of electricity (LCOE). This takes into account the capital cost and the cost of financing the investment together with fuel costs and operations and maintenance costs over the lifetime of the technology. The costs of finance can vary considerably from country to country and also depend on the entity developing the project. Large companies can normally access cheaper finance than small ones, and sovereign states can normally access much cheaper loans. The overall interest rate is, in practice, also dependent on the perceived risk associated with the project and also on the relative proportions of debt and equity and their corresponding interest rates. Capital costs can also vary significantly between different countries and regions. Generation plant require different levels of back up depending on their availability, a factor that also adds a cost. All this makes it difficult to come up with specific figures and most studies show a considerable range of values. Figure 7.1, in which external costs are not accounted for, is taken from a 2017 report from the International Renewable Energy Agency (IRENA), [1], and is based on a 10% annual cost of money (except for OECD countries and China for which 7.5% is assumed). It is clear that, even in the absence of compensation for external costs, renewable sources are now generally competitive with fossil fuels.

7.3 Economic Optimisation in Power Systems

7.3.1 Diversity of Generator Characteristics in a Power System

Power systems are in general fed by a diversity of generators using a range of fuels. Each generator type has its own characteristics and is suited for a particular function within the system. Table 7.2 summarises these characteristics and extends the features covered in Table 2.6 of Chapter 2.

The optimum order in which plant are brought on line to meet load demand at any time, is addressed in the next section.

7.3.2 Optimum Economic Dispatch

One of the aims in power system operation is to load generating sets in response to demand changes so as to minimise the cost of supply. This is known as *Optimum*

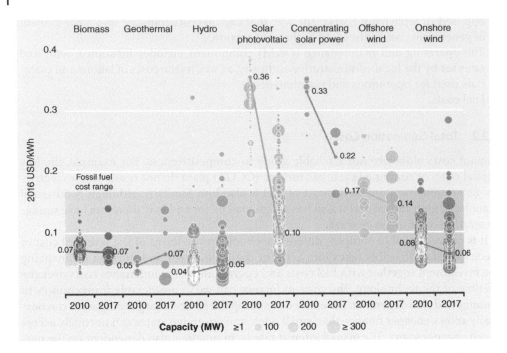

Figure 7.1 Levelised cost of electricity for renewable sources 2010–2017. Source: Taken from [1]. *Note:* The diameter of the circle represents the size of the project, with its centre the value for the cost of each project on the *y*-axis. The thick lines are the global weighted average LCOE value for plants commissioned in each year. The band represents the fossil fuel-fired power generation cost range.

Table 7.2 Operational characteristics of traditional generation.

Type of plant	Physical characteristics	Economic characteristics	Role within system
Nuclear	Inflexible. Best operated at steady load	Very high capital cost. Very low running cost	Steady base load. Average of 75% capacity factor over life time
Gas CCGT	Can be flexible But poor load following response	Low capital cost, low running costs. Fuel supply contracts dictate running on high load	Steady base load. 80% capacity factor
Coal	Flexible with good load following response	Medium capital cost, medium-high running cost	Demand following, units partly loaded to provide reserve. Capacity factor from 20 to 70%
Pumped-storage hydro	Extremely flexible	High capital cost, high marginal cost (see Section 7.3.3)	Rapid response used for 'peak shaving' or in emergencies
Hydro with reservoir	Flexible	High capital cost low running cost	Base load and demand following if required
Open cycle gas turbine	Flexible	Very low capital cost, very high running cost	'peak shaving'. Capacity factor ≪5%

Economic Dispatch (OED) and over the years sophisticated OED methods have been developed to identify the minimum cost solution and at the same time satisfy the numerous engineering and operational requirements and constraints. Here the methods used to adjust the load on individual conventional or RE generators to achieve minimum production cost on a short-term basis will be examined. Such approaches would be applied by the operators of nationally owned power systems. Contemporary market approaches, to be discussed later, try to emulate this behaviour through the structure and operation of a power generation market.

A crude way of applying economic dispatch is to assume that the cost of operation of a plant is a linear function of the plant loading level, that is,

$$C_A = K_A + K'_A P_A$$

where C_A is the cost in £/h for plant A, K_A and K'_A are constants related to the capital and fuel plant cost components respectively and P_A in MW is the power delivered by the plant. To determine the optimum load sharing between two plants A and B when the total demand on the system is P_d the linear cost curve for unit A is plotted normally as in Figure 7.2, while the axis of the unit B curve is rotated by 180°. The origin O_A of unit A is placed at a distance from O_B equal to the total load $P_d = P_A + P_B$.

Any vertical line intersecting the curves will represent the cost of a given operating condition. A sharing of the demand by the two generators is sought such that the cost is minimum. From the diagram, it is obvious that the minimum cost C_{min} occurs when the unit A takes as much of the load share as possible, that is, it is loaded to its maximum level, with unit B brought in to cover the deficit.

The *incremental cost* of a plant is provided by the slope of the cost curve, $dC/dP = K'$. In the example of Figure 7.2, unit A might be a nuclear plant with very large capital costs (hence large K_A) and very low fuel costs (hence, low incremental cost K'_A), while unit B could be a coal-fired station with lower capital costs but higher incremental costs.

A plant loading methodology is now clear. If there are n plants available arrange the plants in order of increasing K', the so-called *merit order*. Starting from the top of the list, the plants are then loaded (each up to the limit of its capacity) before the next plant on the list is brought into action. This loading approach results in the example scheduling shown in Figure 7.3.

To summarise, in the merit order approach, units with low operating costs run preferentially and therefore attract high load factors; they generate a disproportionately large share of electricity relative to their capacity. These high-merit plant are called *base load plant*. Units with high operating costs are only run during peak demand, generate a

Figure 7.2 Demonstration of merit order loading.

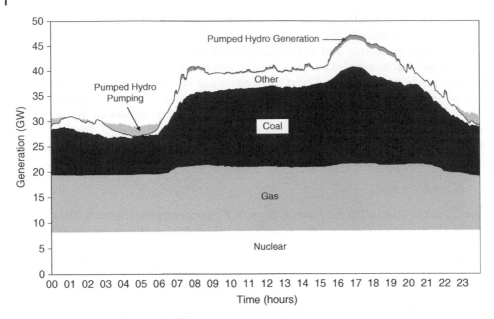

Figure 7.3 Typical generator scheduling to meet demand.

disproportionately small share of the total electricity and are known as *peaking plant.* Plant in between these two extremes is called *intermediate* or middle merit plant.

Figure 7.3 illustrates this loading process. In conventional systems without renewable energy plant, nuclear generation having low operating costs is allocated first. More efficient gas and steam thermal generators are next in merit, and finally fast response gas turbines and hydro pumped storage are used to meet demand during short-term peak periods.

7.3.3 Equal Incremental Cost Dispatch

The merit order philosophy is adequate and convenient for rough scheduling but as a precise tool of economic dispatch it is inaccurate. The reason for this is that cost curves are not exactly linear as assumed previously; this nonlinearity can be approximately described by a quadratic function as in the example of Figure 7.4. Such curves can be used to determine the overall efficiency of the plant at various output levels.

As was done previously, let it be assumed that only two plants A and B are supplying a demand P_d. Following the plotting procedure used in Figure 7.4, the curve for unit A is plotted in Figure 7.5 conventionally while the axis of the unit B curve is rotated by 180°. Again, the origin OB of unit B is placed at a distance from OA equal to the total demand $P_d = P_A + P_B$. As before, any vertical line intersecting the curves will therefore represent the cost of a given operating condition and a sharing of the demand by the two generators is sought such that the cost is minimum.

Let the vertical line C_1 represent the operating point at which the slopes of the two curves are equal. The length of C_1 represents the total cost per hour of operating both units. Any other operating points (such as the ones represented by lines C_2 or C_3) result in higher cost. Hence the most economic operating point is the one at which the slopes

Figure 7.4 Typical input-output curve for a thermal generating unit.

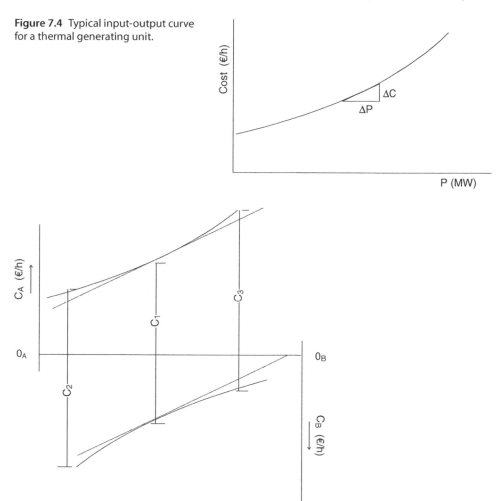

Figure 7.5 Demonstration of the equal incremental cost solution.

of the cost lines (i.e. the *incremental costs*) of the two units are equal. This result is valid no matter how many generators are feeding power to the system. The same conclusion can be proven using the mathematical optimisation technique of Lagrange multipliers. Note that another term for incremental cost is *marginal cost*.

7.3.4 OED with Several Units and Generation Limits

Steam driven generators have not only an upper limit of loading that corresponds to their name plate rating but also a lower limit that is imposed by cavitation that occurs in the turbine if the steam throughput falls below a certain critical level. This lower limit may be anything between 30 and 50% of the generator rating.

Differentiation of quadratic cost curves gives a linear function. Figure 7.6 shows incremental cost curves for a four generator power system (with upper and lower limits for each generator) supplying a demand $P_d = P_A + P_B + P_C + P_D$.

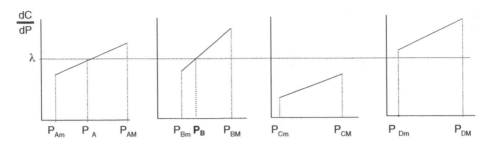

Figure 7.6 Economic distribution among four generators.

It can be seen that the whole range of incremental costs of generator C are below that of the other units, hence C should be loaded up to its full capacity P_{CM}. The new demand $P'_d = P_d - P_{CM}$ has to be distributed among the other three generators according to the equal incremental cost criterion.

Note that unit D, with a higher range of incremental costs, might be shut down (if units A and B are capable of providing the required reserve when partly loaded) or might also be required to provide reserve in which case it will be loaded at its lower limit P_{Dm}. Assume that enough reserve can only be provided if unit D is loaded at its lower limit. The new demand now is

$$P''_d = P'_d - P_{Dm}$$

which has to be optimally shared between units A and B.

Now choose a horizontal line (shown as λ in the figure). Suppose that the height of that line is such that it intercepts the incremental cost curves so that

$$P_A + P_B = P''_d$$

This is the graphical solution to the optimal economic dispatch problem in which the incremental costs of the generators operating between their limits are the same and equal to λ. The practical inadequacy of the merit order scheduling should now be obvious. The difference between the incremental costs of a plant at its minimum and maximum operating levels may exceed the difference between its average operating cost and the average operating costs of the plant on either side of it in the merit order list. The simple merit order technique is therefore an imperfect tool for economic dispatch.

It should be appreciated that even if the total cost of generated energy by a power station, for example a nuclear power station, may be higher than that from a coal-fired one, the OED requires that the nuclear station is fully loaded because of its very low incremental cost. The rationale is that the loading priority is independent of the original capital sum expended in the construction of a power station. As the station is present and available it should be utilised fully because of its low running costs. Whether the station should have been built in the first place is another matter. We may conclude that as the energy supplied by RE sources (except biomass) is characterised almost entirely by the capital cost component, their incremental cost is negligible and hence the OED method would require them to provide at all times the maximum energy they are capable of delivering.

Worked Example 7.1

A small power system is fed from three fossil fuel-fired generators the cost characteristics of which in £/h are

$$C_1 = 500 + 20P_1 + 0.1(P_1)^2$$
$$C_2 = 600 + 24P_2 + 0.15(P_2)^2$$
$$C_3 = 450 + 16P_3 + 0.2(P_3)^2$$

All generators have an upper power limit of 100 MW.

a) Calculate the OED when the load on the system is 250 MW.
b) Hence, calculate the cost in pence per kWh of the generated electricity.

A wind farm providing an average power of 50 MW is now connected to the system.

c) Calculate the new OED.
d) Hence, calculate the cost in pence per kWh of the generated electricity
   Comment on the difference between (b) and (d).

Model Answer

Apply OED based on Lagrangian technique.

$$\frac{dC_1}{dP_1} = 20 + 0.2P_1 = \lambda \qquad (7.1)$$

$$\frac{dC_2}{dP_2} = 24 + 0.3P_2 = \lambda \qquad (7.2)$$

$$\frac{dC_3}{dP_3} = 16 + 0.4P_3 = \lambda \qquad (7.3)$$

$$P_1 + P_2 + P_3 = 250 \qquad (7.4)$$

Substitute Eqs. (7.1)–(7.3) into (7.4).

$$\frac{\lambda - 20}{0.2} + \frac{\lambda - 24}{0.3} + \frac{\lambda - 16}{0.4} = 250$$

$$0.12(\lambda - 20) + 0.08(\lambda - 24) + 0.06(\lambda - 16) = 6$$

and

$$0.26\lambda = 6 + 5.28$$

$$\therefore \lambda = \frac{11.28}{0.26} = 43.58$$

Substituting into (7.1)–(7.3)

$$P_1 = \frac{43.3 - 20}{0.2} = 116.9, \qquad P_2 = 64.6, \qquad P_3 = 68.45$$

But P_1 hits upper limit, hence set $P_1 = 100$ and demand to $250 - 100 = 150$.

a) Apply OED again using Eqs. (7.2) and (7.3) only.

$$\frac{\lambda - 24}{0.3} + \frac{\lambda - 16}{0.4} = 150$$

hence

$$0.4(\lambda - 24) + 0.3(\lambda - 16) = 150 \times 0.12$$

and

$$\lambda = 46.285$$

substituting into (7.2) and (7.3) gives

$$P_2 = \frac{46.785 - 24}{0.3} = 74.28$$

$$P_3 = \frac{46.785 - 16}{0.4} = 75.712$$

b) Substitute P's into equations to find £/h.

$$\Sigma C = 3500 + 3210 + 2867.85 = 9517.85 \ \text{£/h}$$

$$Demand = 100 + 74.28 + 75.712 = 250 \ \text{MW}$$

Hence
Cost in pence per $kWh = 9517.85/250 \times 10^3 = 3.8\text{p/kWh}$

c) With 50 MW wind input, Demand = 250−50 = 250 MW
Hence

$$\frac{\lambda - 20}{0.2} + \frac{\lambda - 24}{0.3} + \frac{\lambda - 16}{0.4} = 200$$

$$\therefore 0.26\lambda = 4.8 + 5.28 \qquad \lambda = 38.769$$

and

$$P_1 = \frac{38.769 - 20}{0.2} = 93.84$$

$$P_2 = \frac{38.769 - 24}{0.3} = 49.23$$

$$P_3 = \frac{38.769 - 16}{0.4} = 56.92$$

d) Cost $C = 3257 + 2145 + 2008 = 7410$
cost per unit $£7410/200 \times 10^3 = 3.7 \ \text{p/kWh}$

e) Renewable energy sources are in general characterised by approximately zero energy cost. As a consequence, they are at the top of the merit order list and their input is simply treated as negative load.
At a reduced demand, some of the less economic generation is either partly loaded or decommissioned, hence the general cost of generating electricity from the conventional sources (in p/kWh) may be reduced.

7.3.5 Costs on a Level Playing Field

Although generation costs are commonly used to compare renewable energy and fossil generation, this process does not account for all factors and so is not precise. A so-called 'level playing field' demands that allowances are made for all relevant factors, some of which add to and others subtract value from renewable power generation. More accurately, the benefit of a new renewable source of energy for an electricity network may be specified as: (value due to fuel savings) + (value due to reduction in external costs) ± (embedded generation benefits/disbenefits) + (value of capacity credit) − (costs associated with variability), where:

- The *fuel saving value* is simply the value of fuel saved for each MWh generated. In the UK in 2017, this amounted to about £30/MWh. The value varies on an hourly basis so this is a rough estimate of the annual average. The concept of capacity credit and the penalties due to variability were discussed at length in Chapter 3.
- *External costs* are costs attributable to an activity that are not borne by the party involved in that activity. All electricity-generating technologies have associated external costs, and those of the fossil fuel sources of generation are due to the pollution that arises from their use, and from the impacts of global warming due to their CO_2 emissions. Economists argue that these costs should be added to the generating costs, and this is the thinking behind carbon taxes. However, some consider that these would add unacceptable increases to the price of electricity and so most governments prefer to give renewable energy sources some subsidy instead.
- *Embedded generation benefits/disbenefits* acknowledge that many renewable energy sources are small-scale and so connect into low voltage distribution networks. This means that losses in the electricity network may be reduced and, possibly, transmission and/or distribution network reinforcements delayed or deferred. The calculation of these benefits is a complex issue, and varies both regionally and locally. It is important to recognise, however, that, as mentioned in Chapter 6, concentrations of embedded generation can increase distribution losses in rural areas where demand is low and so may result in disbenefits.
- Value of *capacity credit* and extra *balancing costs* due to the RE source variability. These topics were introduced in Chapter 4.

All of these key components are discussed in the following sections.

7.4 External Costs

7.4.1 Introduction

Although there is general agreement as to the broad definition of external costs being costs attributable to an activity that are not borne by the party involved in that activity, there are widespread variations in how the boundaries are defined. For example, some have proposed that a substantial proportion of Western defence budgets should be regarded as an 'external cost' of oil, and in light of recent conflicts, this may not be wholly unreasonable.

External costs can be difficult to quantify but they can be of significant magnitude. If the enormous costs of the clean-up operation after the Chernobyl and Fukushima nuclear disasters had been taken into account when these stations were planned, they clearly would never have been built. External costs are set to play an increasing role in shaping future energy policy. Cleaning up the business of power supply is now a fundamental aim of many governments around the world, and in many countries, it is now embedded firmly in energy policy. It is now recognised that the costs to society of pollution is substantial, costs to which the electricity industry is self-evidently a major contributor. The task facing energy policy makers is how best to go about the job of reducing pollution from electricity generation when in most countries external costs are not reflected in the market price of the end product. If this was done in a realistic manner, fossil fuel technology and nuclear prices would rise significantly, making renewable energy sources more competitive.

Deregulation of the power markets can either aid or hamper the quest for a proper recognition of external costs depending on the details. Crucial to the process is the willingness of governments to mandate that all energy options should compete on an equal footing. This would encourage utilities to take the full costs of electricity supply into account when planning new generation capacity. Such levelling of markets will also force the hidden subsidies attached to conventional technologies into the open; this *unbundling* is a primary aim of privatisation, or deregulation. Overall, what is needed are fair and workable procedures to achieve what is referred to as the *internalisation* of external costs.

7.4.2 Types of External Cost

Before looking more closely at the procedures available, a brief analysis of the make-up of external costs will help explain why they engender controversy. Table 7.3 gives a list of the local and general external costs excluding global warming.

External costs associated with energy supply/use are complicated. For simplification they can be divided into three broad categories.

1. hidden costs borne by governments
2. costs of the damage caused to health and the environment by emissions other than CO_2
3. the costs of global warming attributable to CO_2 emissions.

The first category includes the cost of regulatory bodies and pollution inspectorates (generally small) and the cost of energy industry subsidies and research and

Table 7.3 Sources of external cost due to electricity production, excluding global warming.

Local	General
Smut deposition	Acid rain damage
Obscuration of the sun by plumes	• to trees and crops
Noise – due to plant, coal handling, etc.	• to buildings
Noise due to fuel deliveries	• to human health
Discharges into watercourses	• to fisheries
Plant accidents – human cost	• to animals
Smells	Oil spillages – clean-up costs
Dust and fumes	Ash disposal accidents
Heavy metal depositions	Heavy metal depositions
Upkeep of emergency evacuation measures (Nuclear)	Leakage from radioactive waste

development programmes. These can be very significant. In the ground-breaking analysis of external costs published by the European Commission in 1988 [2], it was calculated that support to the German coal industry added DEM (German Deutsche Mark) 0.002/kWh to the price of electricity. A cost of DEM 0.0235/kWh was also assigned to nuclear research and development, compared to around DEM 0.004/kWh for wind power. Despite EU aims to stop support to fossil fuels by 2020, Germany still subsidises coal to a significant extent. Between 2015 and 2016, this was estimated by the Federal Ministry of Finance (BMF) to be at a rate of €1.2 billion per year for sale of coal for electricity generation.

The second category covers costs due to emissions that cause damage to the environment and/or people. These make up about 10% of the external cost of power generation and include a wide variety of effects, including damage from acid rain and health damage from oxides of sulfide and nitrogen emitted from coal-fired power stations. In the European study, ExterneE, the costs of damage to health were estimated by calculating the loss of earnings and costs of hospitalisation of people susceptible to respiratory diseases.

Other costs included in the damage and health category are power industry accidents, whether they occur in coalmines, on offshore oil or gas rigs, or in nuclear plant. The probability of a serious nuclear accident in Western Europe might be extremely low, but should a catastrophic failure occur the costs would be undeniably huge.

The third category is by far the largest: external costs due to greenhouse gas emissions, which cause global warming with all its associated effects. This category accounts for some 40–100% of the hidden costs of the world's consumption of electricity. It is also the most contentious area of the external costs debate. The range of estimates for the possible economic implications of global warming is extremely large. Costs associated with climate change, flooding, changes in agricultural patterns and other effects all need to be taken into account. Although calculating the costs of greenhouse gas emissions is problematic, there is a degree of agreement about the magnitude (or at least the range, since local factors affect the numbers themselves) of such emissions associated with different generation technologies. It is conventional to present these emissions as whole lifecycle emissions including all greenhouse gases as equivalent CO_2, thus taking into account embodied energy of manufacture, recycling, and other indirect emissions. Figure 7.7 is taken from a recent InterGovernmental Panel on Climate Change IPCC report [3]. It is interesting to note that even with Carbon Capture and Storage (CCS), coal and gas-fired electricity generation still produce more emissions than the renewables.

7.4.3 The Kyoto Protocol and Subsequent Agreements

Most high-profile global agreements to curb man-made emissions of greenhouse gases, widely believed to be contributing to global warming, stem from the 1997 Kyoto Conference. The agreements emerging from this conference are obligatory for ratifying countries through the United Nations Framework Convention on Climate Change (UNFCCC), with associated penalties for non-conformance. These agreements formed the basis for reducing the emissions of a 'basket' of greenhouse gases from the industrialised countries (referred to as 'Annex 1', or 'Annex B' countries) by 5.2% of the 1990 levels by a commitment period between 2008 and 2012. It was expected that this reduction would be achieved in each Annex 1 country through the 'domestic

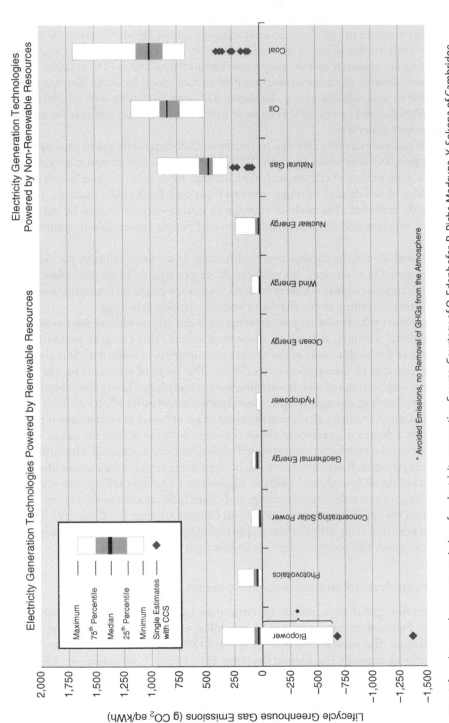

Figure 7.7 Lifecycle greenhouse gas emissions for electricity generation Source: Courtesy of O. Edenhofer, R. Pichs-Madruga, Y. Sokona of Cambridge University Press [3].

measures' of introduction of more efficient, and/or less carbon intensive, systems of power generation and industrial processing, with possibilities for the development and management of forests and agricultural soils ('carbon sinks') to absorb carbon dioxide from the atmosphere. These 'domestic measures' were supported by the use of international 'flexible measures' such as emissions trading, joint implementation (JI), and the clean development mechanism (CDM). Countries signed up to Kyoto have introduced a range of policy measures aimed to ensure they met their emissions targets. Some of these have the effect of internalising at least some of the external costs associated with global warming. In December 2012 in Doha, it was agreed to extend the Kyoto Protocol until a new international agreement could be put in place for 2020. The world's two largest emitters of greenhouse gases, the United States of America and China, remained as non-signatories. The most significant recent development that builds on the Convention is the 2015 Paris Agreement by which signatories agreed to limit global temperature rise this century to well below 2 °C above pre-industrial levels, and to pursue efforts to limit this warming to 1.5 °C. The agreement came into force on 4 November 2016, but significantly the USA withdrew from the agreement in 2017. Every five years it is intended to undertake a quantitative assessment of progress made. The primary weakness of the agreement is that there are no legally binding constraints on the signatories and no hard national targets for emissions. In this important sense, international policy to deal with global warming is now less committed than at the time of the Kyoto accord.

7.4.4 Costing Pollution

Costing pollution is not easy and can be problematic so devising market mechanisms that take external costs into account is a challenge. The approach favoured by economists over recent years has been *carbon taxes*. Such a tax is directly linked to the level of carbon dioxide (CO_2) emissions, often expressed as a value per tonne of CO_2 equivalent. This sends a price signal to the energy market that discourages the use of fuels with high associated CO_2 emissions. In the case of electricity, it simply adds to the cost of generation using fossil fuels in a manner that reflects the relevant external costs due to their CO_2 emissions. Many countries, including the UK, Denmark, Finland, Sweden, the Netherlands, France, Japan and India, have implemented such taxes although sometimes these only apply to specific fossil fuels. The EU as a whole has not mandated such taxes; instead it introduced a system of *carbon trading* that commenced in 2005. The EU Emissions Trading System (ETS) sets limits on member states' carbon emissions from large industrial plants including fossil fuelled power stations. The intention of the ETS is that the limits for each country will be steadily reduced over a number of years to help deliver member states' obligations under the Kyoto Agreements. In practice, fossil fuelled power plants have an annual quota for carbon emissions and those who exceed their allowances must purchase 'credits' from generators or others who have not exceeded their allowances. In February 2005, allowances traded at just under €10/tonne of CO_2, which corresponds to a premium of €1/MWh on the price of coal-fired generation, and about half that amount on gas-fired generation. However, in January 2013 the value of EU allowances fell to €2.8/tonne and there is now increasing concern about the viability of carbon trading.

Another approach to external costs, favoured by some American states, is to use integrated resource planning (IRP) to assess future generating plant needs. IRP is a process whereby the different options for meeting future electricity demand are considered and the least cost solution for delivering energy is selected within the constraints imposed by ensuring security of supply. The benefit of IRP is that, in complying with government mandates to consider all energy supply options, the regulatory body simply assumes that external costs are to be applied, without actually imposing any. The precise monetary value of pollution damage, or the calculation of cost penalties on polluting fuels, becomes less critical. This straightforward approach does away with many of the complicated mechanisms required to calculate the monetary value of pollution in electricity generation. It also helps to ensure that power plants are built in the right place and that the electricity network operates efficiently. The result is a competitive market operating under guidelines aimed at long-term economic responsibility. However, this approach has not been without problems. Recent studies in California and elsewhere indicate that the net result of this approach is simply to shift the utility plant mixes in favour of gas, which is currently cheap and has lower external costs than coal. Renewables profit, but the additional market share assigned to them so far remains relatively small. It is important to remember, however, that gas is unlikely to remain cheap indefinitely. In addition, government and popular pressure will accelerate the introduction of renewable energy technologies.

Popular consumer desire for more clean electricity is illustrated in the trend towards *green pricing* in the US, Australia and Europe. Given the choice, electricity consumers often agree to pay more for green electricity. In the Pacific Northwest of the US, a modest 4% increase in rates is making a utility 20% 'green' and the initial reaction of the consumers is 75% in favour. But while green pricing is a clear signal to governments to promote renewables in the electricity market, it cannot be regarded as a substitute for an equitable energy policy. Furthermore, green pricing will not necessarily deliver the most efficient system overall. It is likely that a degree of IRP would need to be integrated into the market.

There are various other ways of reflecting external cost, directly or indirectly. Market subsidies are often justified as in some way reflecting the external costs, but they can also play the role of encouraging new technologies until they become competitive. The various forms that subsidies can take are discussed in detail Section 7.6. It should be noted that support mechanisms historically have changed on a frequent basis, reflecting electricity markets evolution and the development of the renewable energy sector. Reference [4] discusses external costs associated with electricity generation in some detail.

7.5 Effects of Embedded Generation

7.5.1 Value of Energy At Various Points of the Network

In the case of small-scale renewables there is the potential for an *embedded generation benefit*. A simple example may be used to illustrate the point. If a householder were to install a photovoltaic (PV) panel on the roof of her/his house and export the surplus electricity to the local utility, they would resell it at between €13 and €15 cents/kWh to neighbours. There would be no backflow through the local 33/11 kV transformer and, as a bonus, the locality might suffer fewer power cuts.

This reasoning may be simplistic but it is nevertheless the principle behind *net metering*, which is used by several US utilities to remunerate renewable energy generators at 100% of the supply cost. A modified version of the approach has been used in Denmark and Germany where renewable generators were paid 80–90% of the domestic tariff.

Although studies in the US and elsewhere point to the fact that renewable energy may have a high value in certain locations, this point is not widely accepted within EU utilities, many of whom continue to assign renewable energy a value based only on its fuel saving capability. The cost, price and hence the value of electricity varies on a geographical basis. Above-average costs are incurred when energy has to be transmitted long distances from the generating plant. The need to supply remote locations increases the costs of the distribution network and, in addition, electrical losses are incurred in feeding the energy to the extremities of the system. In many instances, renewable electricity-generating technologies deliver energy closer to consumer demand than centralised generation. It substitutes for electricity that has accrued a higher selling price than the generation costs of large thermal plant, due to its passage through the network. This rising level of 'value' with reducing voltage is reflected in higher charges for customers connected at lower voltage levels. This variation in value of electricity is illustrated in Figure 7.8, which includes typical selling prices for electricity at various voltage levels in the UK system for 2018.

Electricity utilities rarely ascribe exact costs to account for all geographical and temporal factors. Tariffs build in cost averaging so that blocks of consumers with similar requirements and locations pay similar rates.

7.5.2 An Example Cash-Flow Analysis

As part of an EU-funded study [5], an analysis of the cash flows through the electricity network in England and Wales was carried out. Although the study itself is now rather

Electricity distribution and renewable energy

Prices are average levels in € cents/kWh, and are approximate

Figure 7.8 Changing electricity value through the power system from bulk generation to consumers.

dated, the broad conclusions still apply. Data were compared with information from tariffs, and with other case studies in Europe and the United States. The analysis enabled average levels of electricity price to be established at key points in the distribution chain:

- at the busbars of large thermal plant
- at exit from the high voltage transmission system
- at delivery to the consumer

Conventional terminology delineates three stages of electricity supply:

- generation
- high voltage transmission (often abbreviated to 'transmission')
- distribution

For each of these stages in the supply chain, costs can be allocated under four headings:

- *fuel* (fuel cost purchases for generators and power losses in transmission and distribution)
- *operating* reflecting all operating costs including overheads
- *depreciation* of capital assets
- *profit* on a historic cost basis

The cash-flow analysis was carried out for 1993/1994 and costs in nominal money values have remained similar since that time (which means they have fallen in real terms). The results of the cash-flow analysis are summarised in Table 7.4. Although the actual values presented in the table are now considerably out of date, the ratios between them will have remained roughly constant. It is thus still a good guide to how the final cost of electricity is made up.

The entries for 'fuel' under transmission and distribution cover system losses. The price elements shown in Table 7.4 apply to the system as a whole. Regional variations occur for many reasons, primarily due to the proximity of a locality to sources of generation.

Although the constituent elements of electricity price shown in the table, define its 'value' at any given point in the system in average terms, it does not follow that electricity, which is injected at that same point, will be credited with that same value by the generator or utility. A brief summary of current attitudes to value is as follows:

Table 7.4 Segregation of costs in UK ESI, 1993/1994.

	Fuel	Operating	Depreciation	Profit	Total
Generation	2.03	1.28	0.35	0.35	**4.01**
Control		0.14	0.01	0.03	**0.18**
Transmission	0.08	0.11	0.16	0.17	**0.52**
Distribution	0.31	0.80	0.42	0.58	**2.11**
Supply		0.26	0.01	0.13	**0.40**
Total	**2.42**	**2.59**	**0.95**	**1.26**	7.22

Data in € cents/kWh.
Source: Courtesy of David Milborrow.

- *Fuel.* Most studies accept that fuel savings are a component of the value of RE. With variable sources, the value declines as the amount of variable capacity increases.
- *Operating costs.* These include 'Fixed' and 'Variable' components. The variable costs of generation are part of the value of RE, but the variable costs of transmission and distribution are more contentious issues. Acceptance of the *fixed* component of these costs as a component of value depends on whether it is accepted that RE, particularly the variable sources has a 'capacity credit' as discussed in Chapter 3.
- *Depreciation.* The crucial point is whether the introduction of RE on to a system enables plant savings to be made. Most studies have addressed generation plant savings, but more recent work addresses transmission and distribution savings.
- *Profit.* The question of profit as an element of value has not explicitly been addressed. In a deregulated system where no single generator or distributor has exclusive rights, it would seem valid to allow its inclusion, at least partially, in value assessments.

7.5.3 Value of Embedded Generation – Regional and Local Issues

Transmission limitations may impose constraints on the assimilation of new generation capacity, whether conventional or renewable. For example, in the case of wind in the UK, concentrations of wind in Scotland and the North of England, where wind speeds are higher than those further south, may cause an increase in the north-south power flows, which are already substantial. Figure 7.9, taken from [6], shows how the hourly Scotland to England border flow varies with Scottish wind power production. Considerable reinforcement costs for the transmission networks would then be needed, depending on the timing and location of new renewable generation.

Studies carried out indicate that by 2020 several GW of additional north-south transmission capacity will be required for the UK. This can be delivered by a combination upgrades to the existing interconnector by the installation of series capacitors and the installation of High Voltage Direct Current (HVDC) links, the first of which was commissioned in 2016: the so-called Western HVDC link. A second HVDC connector is under consideration. An analysis undertaken by Sinclair Knight Mertz [7] examined

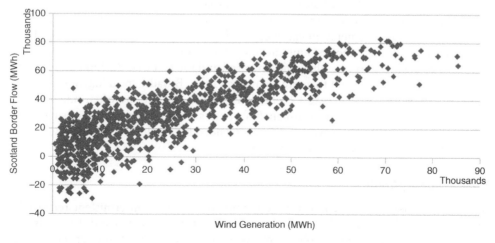

Figure 7.9 Scotland to England cross border flows as a function of wind power generation in Scotland.

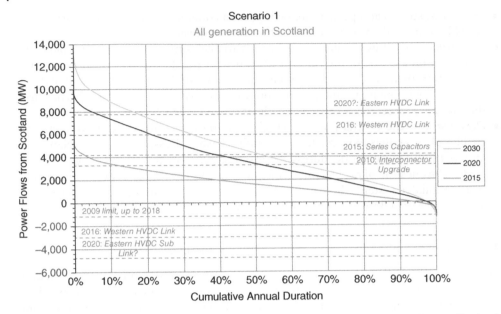

Figure 7.10 Duration curves for power flows from Scotland for 2015, 2020 and 2030. Source: Obtained from Scottish Generation Scenarios and Power Flows: An Analysis, Sinclair Knight Merz, 2011 [7].

the likelihood of such power flows in 2015, 2020 and 2030 in relation to the installed capacity of the interconnector. Figure 7.10 shows the results; the projection for 2020 indicates that the north-south power flow will exceed the network capacity, even with the Eastern HVDC link in place.

In addition to any connection charges specific to the installation of generation, in the UK Transmission Network Use of System (TNUoS) charges recover the cost of installing and maintaining the transmission system. Tariffs are based on geographical zones to discourage new generation remote from load centres and requiring substantial transmission capacity. In practice, higher transmission charges in the North of England and Scotland may dampen the enthusiasm for wind projects.

Local issues are a complex topic as they vary both regionally and locally. Concentrations of embedded generation can increase distribution losses in rural areas where demand is low, although modest amounts may reduce losses.

In view of the complexity of the issues involved and the fact that the impact is dependent on location, it is difficult to quote definitive values. Broadly speaking, however, the embedded benefits rarely exceed €2 cents/kWh in the UK and are often much lower. Renewable generation may incur disbenefits where substantial quantities of such generation, for example, in Scotland may trigger the need for significant (and costly) transmission reinforcements as discussed previously.

7.5.4 Capacity Credit

An important issue in the context of economic appraisals of intermittent renewable sources is capacity credit. This topic was introduced in Chapter 3 and it is revisited here in terms of its economic implications. As mentioned previously the capacity credit of any power plant can be viewed as a measure of the ability of the plant to contribute to the peak demands of a power system. More precisely it is defined as the capacity of

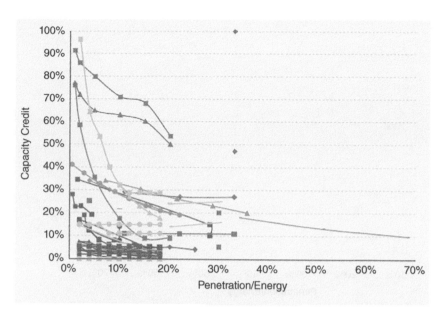

Figure 7.11 Comparison of numerous studies of capacity credit for wind. (UK/IRE are dark circles and dots; EUR North/Central are dark squares and dark diamonds; EUR South are light squares and light circles; US are medium squares and circles).

conventional thermal plant that an intermittent renewable generator can replace while still delivering the same reliability of supply to energy users. Of the renewable energy sources, wind has received the most attention in this regard, with numerous utility studies concluding that wind can displace significant amounts of thermal plant. It is important to note that capacity credit depends not only on the capacity factor for wind, so, for example, offshore wind will have a higher value than wind onshore all other factors being equal, but importantly on the characteristics of the power system and the make-up of the conventional generation. The capacity credit of renewable plant in general will decline with its capacity or level of penetration into the power system. Studies show considerable variations in the calculated values reflecting in the main differing power system characteristics, but also, to an extent, different calculation methodologies.

A recent review of past studies undertaken by the UK Energy Research Centre (UKERC) [8], covering the UK, Europe and the USA shows widely varied estimates of wind capacity credit as a function of penetration level. Figure 7.11 is taken from this review.

Capacity credit will depend also on the correlation between renewable generation and the load. For the case of wind in the UK, for example, there can be expected to be periods of calm during the summer, but the loss of load probability at such times is low. The non-availability of wind during such periods therefore makes little difference to the year-round probability. It may be noted that the capacity credit of solar plant in the UK would be very low, and that of tidal barrage plant would be lower than the average capacity factor. Figure 7.12, also taken from the UKERC study, contrasts the capacity credit values for wind and photovoltaics and also solar thermal power with storage (the two outlier points at 33% penetration).

In the USA, solar thermal power stations have also been assessed for their capacity credit and how this varies with the amount of thermal storage included. Figure 7.13,

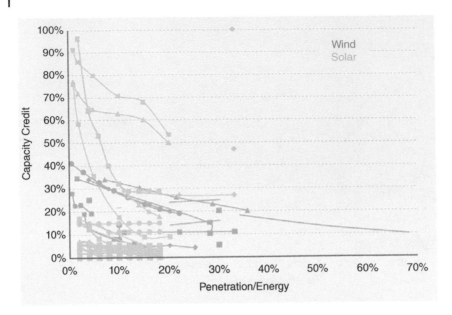

Figure 7.12 Contrasting capacity credit values for wind and photovoltaics.

Figure 7.13 Capacity credit calculations for a solar thermal power plant located in Nevada, USA, for different amounts of thermal energy storage and different solar multiples.

taken from [9], shows the capacity credit (or value) for a single solar thermal plant located in Nevada, USA. Solar thermal storage can be integrated into such plant and, as shown, can enhance the capacity value. Another factor that affects the capacity credit of such plant is the solar multiple, which relates to the maximum capacity of the heliostat to the rated heat transfer and power generation capacity of the solar tower itself.

Once the capacity credit has been determined, the value of capacity can then be estimated on the basis of the capital investment costs for the displaced conventional plant.

7.5.5 Summary

Much attention has been given to the costs of renewable power generation. In recent years, the cost of energy from wind, both onshore and offshore, and photovoltaics has fallen dramatically, to the extent that in locations with a good resource, subsidies are now hardly required. So, whereas in 2003 onshore wind energy supply was estimated to add about 0.3p per kWh (€0.45 cents/kWh) to consumer electricity bills [10], this cost can now be considered to be effectively zero. Indeed, many energy analyses now see renewable power generation as a means to mitigate against expected future increases in the cost of gas-fired generation, and this is without any consideration of the external costs.

7.6 Support Mechanisms for Renewable Energy

7.6.1 Introduction

The previous section assessed the value of renewable energy generation when competing directly with other 'conventional' power generation sources in an open electricity market. Many countries have recognised that renewable energy would find it difficult to compete on level terms due to the issues of:

- *Variability.* As discussed, most renewable energy sources cannot easily be controlled and it is not possible to forecast their output with 100% accuracy.
- *Cost.* Many of the renewable energy technologies are not yet fully mature and thus may cost more than conventional generation at present.
- *External costs.* The external costs of conventional generation (pollution, CO_2 emissions etc.) are not yet properly accounted for and the flip-side of this is that renewable energy generation is not properly credited for avoiding these external costs.

The basic ways that governments have dealt with this situation in order to promote the uptake of renewable energy are:

- The introduction of a *feed-in law or tariff* where different renewable energy generation technologies are paid a fixed price for their output and suppliers/distributors are obliged to take that output.
- A *quota system* whereby suppliers/distributors are obliged to source a certain fraction of their energy from renewable energy sources. This is facilitated by the issuing of green certificates to renewable energy generators that can be traded.
- A *carbon tax* on conventional energy generation, making renewable energy generation more attractive.
- *Carbon trading* (as described before).
- *Tax relief* on the investment in renewable energy generation plant.

This section will discuss the implementation of these schemes by looking at two European countries, namely the UK and Germany, who have implemented all of these measures in some way at some time. An analysis is then made of the value of green certificates under a quota system followed by a brief discussion of the selling of green electricity as a 'product'.

Table 7.5 Tariffs paid to renewable energy generators under the renewable energy sources act as of 2009.

Technology	Tariff (€ cent/kWh)
Biomass <150 kW	11.55
Biomass 150–500 kW	9.09
Biomass 500 kW–5 MW	8.17
Geothermal <10 MW	15.84
Geothermal >20 MW	10.39
Onshore wind (first five years)	9.11
Onshore wind (after five years)	4.97
Offshore wind (first nine years)	13.00
Offshore wind (after nine years)	3.50
Solar photovoltaics (freestanding)	28.43
Solar photovoltaics (roof mounted) <30 kW	39.14
Solar photovoltaics (roof mounted) 30–100 kW	37.23
Solar photovoltaics (roof mounted) 100 kW–1 MW	35.23
Solar photovoltaics (roof mounted) over 1 MW	29.37
Hydro <500 kW	12.67
Hydro 0.5–2 MW	8.65
Hydro 2–5 MW	7.65

Source: Data Courtesy of Couture et al. [11].

7.6.2 Feed-in Law

Germany, like the UK set a target back in 1998 for 10% of electricity to be generated from renewable energy sources by 2010. Since 1991, a law has been in force in Germany that gives a legal obligation for utilities to take off all electricity generated by renewable energy generators at a fixed price depending on the technology. This law was subsequently amended such that the local grid operator took on this obligation instead of the local utility. Feed-in tariffs (generally fixed for 20 years) were set to reflect that state of development and the prevailing generation costs of the different renewable generation technologies; so, for example, a well-developed technology like onshore wind received a lower per-kWh price than solar PV or tidal power. The tariffs were adjusted on a regular basis to reflect technology development. As an example, Table 7.5 shows the tariffs payable to renewable energy generators in 2009 taken from Reference [11]. In 2016, German law was revised to reflect a switch from feed-in tariffs to auctions as the main policy for larger renewable energy projects, although feed-in tariffs remain in place as of 2019 for solar and wind below 750 kW.

7.6.3 Quota System

7.6.3.1 Renewables Obligation (RO)

First introduced in the UK in 2002, and most recently amended in 2015, the Renewables Obligation (RO) places an obligation on all electricity suppliers to source a certain fraction of their demand from renewable energy sources. At the outset, this fraction was 3% and rose to 10% by 2010/2011. If a supplier cannot or will not comply then it can 'buy-out' its obligation by paying at a prescribed rate for the required generation shortfall. This effectively fixes the price for the premium paid to renewable energy generators through a system of certificates known as Renewables Obligation Certificates (ROCs). The proceeds from the buy-out 'pot' are given back to compliant suppliers in proportion to their degree of compliance; that is, how many ROCs they have bought. This can add extra value to the ROCs. ROCs are a source of revenue to renewable energy generators on top of what they are able to receive through trading their output in the wholesale market.

7.6.3.2 Contract for Difference (CFD)

As a key part of the UK's Electricity Market Reform (EMR), from 2015 for projects over 300 MW, the RO was effectively replaced by a scheme of Contracts for Differences (CFDs). A CFD is a private contract between the renewable energy generator and a government owned company: the Low Carbon Contracts Company whereby the generator is paid the difference between a 'strike price' determined by government to reflect the costs of specific low carbon technology and a 'reference price' that reflects the average market price for UK electricity. Consumers are protected from over-payment when the market price exceeds the strike price, since in these circumstances the generator is required to pay back the difference. Generation quotas are not exactly fixed but generation capacity supported by CFD reflects the funding allocation made available to specific technology pots.

7.6.4 Carbon Tax

7.6.4.1 Climate Change Levy

In 2001, the UK government introduced the Climate Change Levy, a form of carbon tax. It is a levy on all energy sold to industrial/commercial consumers and also applies to the agricultural sector and public services providers. For electricity the rate from April 2019 will be £8.47/MWh. Electricity from renewable energy sources was exempt from this levy up until August 2015. The exemption scheme was run using a Levy Exemption Certificate (LEC) scheme. Each certificate had a stated value that could be passed on in part or full from the supplier to the generator for the electricity that the supplier supplies to its commercial customers. This levy can be considered as a type of carbon tax, though electricity from nuclear and large-scale hydro generation, and now renewable generation are *not* exempt. The revenue from the tax is recycled in order to reduce employer national insurance contributions, thus effectively reducing tax on labour to compensate. In their day, LECs were also an additional source of revenue over and above the value of the electricity traded by the renewable energy generator competitively.

7.6.4.2 Eco-Tax Reform

On the 1 April 1999, a tax was introduced in Germany on energy use in order to promote energy efficiency and renewable energy sources. The initial value of the tax on electricity was €10.20/MWh rising by €2.60/MWh annually; following substantial revision in 2014 to support Germany's transition from nuclear power generation the value has risen and in 2018 was €68.8/MWh. In 2018, the tax as a whole generated a record total value of €41 billion. As in the UK, this effective carbon tax is used to offset labour tax by reducing statutory pensions insurance paid by employers and employees. Reductions are allowed for certain domestic, commercial and industrial users. Electricity from renewable energy generation sources is exempt from the tax.

7.6.4.3 Tax Relief

To promote the building of Combined Heat and Power (CHP) plants, wind farms and biogas plants, German investors are able to offset a certain fraction (10%) of the investment costs over 10 years against income tax. This basically gives tax relief on the building of certain renewable energy generation plants.

7.7 Electricity Markets

7.7.1 Introduction

This section will provide an insight into the operation and structure of a modern electricity market, highlighting the differences between a centrally planned and operated system through to a fully deregulated system. This background will stress the challenges faced by renewable energy generation when trying to compete in modern day markets. In addition, the concept of 'green electricity' trading will be introduced drawing on the twin strands of: (i) government legislation to promote the uptake of renewable energy and (ii) industry marketing to sell green electricity as a product in demand in the consumer marketplace.

An overview of electricity trading in several countries will be provided but the UK experience will be highlighted as an example of a system that has pioneered electricity deregulation and has seen a market for green electricity develop through this deregulation. The UK example also illustrates the difficulties and evolutionary processes necessary in ensuring that a fully deregulated system runs efficiently and securely.

7.7.2 The UK Electricity Supply Industry

7.7.2.1 The State-Owned Central Electricity-Generating Board

Before 1990, the entire UK electricity supply industry (ESI) was in the state sector and was centrally planned and operated by the Central Electricity-Generating Board (CEGB). The remit of the CEGB was to provide a secure supply of electricity to its consumers. All generating plant, transmission lines, distribution lines, substations and metering were owned and operated by the CEGB as a monopoly. During the days of the CEGB, electricity was supplied to consumers through Regional Electricity Companies (RECs) that were also responsible for the operation of the distribution network in their area (132 kV and below).

Although this system had worked well for many years, it was a monopoly and severely restricted opportunities for small-scale privately owned generation. During privatisation, that was intended to promote competition and drive down costs, the CEGB was broken up into various sectors and sold off to the private sector and new market mechanisms were brought into being.

7.7.2.2 The Electricity Pool

In 1990, the Pooling and Settlement Agreement (PSA) was introduced and the ESI privatised. The generators were split up between two privatised companies: National Power and Powergen but nuclear power stations remained under state control. The RECs were individually privatised and given sole licences to supply electricity in their areas still retaining responsibility for the distribution network in their areas. The RECs jointly owned the National Grid Company (NGC), which was responsible for the operation of the transmission network (275 and 400 kV) and the pump-storage facilities in Wales. NGC also served as the network operator responsible for the scheduling and dispatch of the generation plant.

7.7.2.3 The Operation of the Pool and Pool Rules

Instead of generation plant being scheduled centrally in an OED mode reflecting efficiency, operating cost and flexibility, generators now 'bid' into a pool offering to generate a given output at a given price for the day ahead. NGC would examine the bids, put them in order of increasing cost and then from this list select the generation plants it would need to meet the demand in a given half-hour period, allowing for transmission constraints and reserve. The most expensive plant required in each half-hour period would set the so-called System Marginal Price (SMP) for that half-hour. To this price would be added a capacity credit in order to ensure that sufficient capacity over and above maximum demand would always be available for reasons of security of supply. This price was called the Pool Purchase Price (PPP) and was paid to *all* generators selected within the half-hour regardless of whether they had bid less. This pool system is known as a *uniform auction*.

Inflexible plant such as nuclear would commonly bid in at a very low price or even zero to ensure being selected. Note that they would still be paid the PPP even if they had bid zero. CCGTs with fixed gas contracts would also bid low to ensure being scheduled. The next lowest prices would be operated by the coal-fired generators that were more flexible but with higher operating costs. Finally, open cycle gas turbines and pumped storage would bid higher due to higher operating costs and greater flexibility. It can be seen that, in general, the order of the bids would be in line with the old merit order system but now prices were meant to be driven down due to the competitive bidding nature of the Pool.

Figure 7.14 shows a comparison of the monthly average daily half-hourly demand (in MW, right y-axis) and the PPP (in €/MWH, on the left y-axis) for November 1999. Although almost 20 years old this data is presented as it illustrates well the working of this market system. It can be seen that the trend in prices throughout the day broadly follows the demand trend during the day. The peaks in prices reflect the more expensive plant that were chosen by NGC to meet the peaks in demand.

The PPP was then 'uplifted' slightly to cover the difference in the total cost incurred by NGC in balancing generation and demand and what was paid out to generators through the PPP. This uplifted price was known as the Pool Selling Price (PSP). The PSP is then charged through to suppliers for the amount of electricity supplied to their customers.

All generation and demand is metered. Large generators and consumers are metered every half-hour. Small consumers are metered monthly, quarterly or six-monthly. All meter readings are processed and made available to the System Operator (SO) who works out who should be paid what amount and who owes what amount. Additionally, charges are made for the use of the transmission and distribution networks. The process of allocating the costs and payments is known as *settlement*.

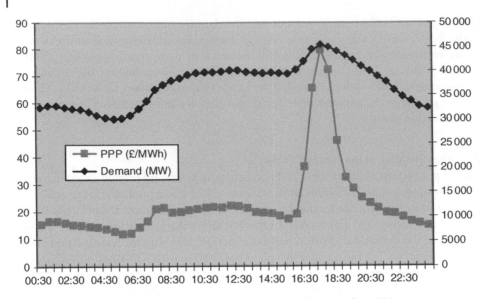

Figure 7.14 NGC average daily half-hourly demand and PPP for November 1999.

7.7.2.4 Hedging

It can be seen from Figure 7.14 that there is a wide variation in prices during the day. The prices shown are monthly averages and thus hide considerable price volatility. It was quite possible to get price fluctuations at times of peak demand (around 5:30 p.m.) well over £100/MWh (€150/MWh). In certain severe weather conditions this could go over £250/MWh (€375/MWh). For a supplier with a peak 'tea-time' demand of 2 GW (2000 MW) this could mean half-hourly cost fluctuations approaching £0.25 million (€0.38 million)! Most suppliers operate to quite tight margins and charge a fixed price to their customers so a sudden increase in Pool prices over a cold period could potentially bankrupt a supplier. On the other hand, generators are paying a fixed price for their fuel and similarly a dip in prices during a prolonged warm period could also spell disaster for their cash flow. Soon after the introduction of the PSA, so-called 'Contracts for Differences' came into existence. These 'financial instruments' gave generators and suppliers the ability to fix the price of their generation or demand for a given period of time, typically a month or more, but also up to several years ahead. The contracts did this by *hedging* against changes in the Pool prices.

Typically, a supplier would forecast its demand profile by half-hour throughout a month taking into account changes due the seasons, changes due to expected customer losses/gains and so on. This would be the *hedged volume*. The supplier would present this profile to a third-party broker, which could be a financial institution. This broker would then offer a fixed price for the contract reflecting how volatile the broker expected the prices to be in the next month. The supplier would still have to pay the half-hourly PSP for demand but at the same time if the PSP went above the fixed price, the broker would pay the supplier for the difference between the PSP and fixed price multiplied by the hedged volume. On the other hand, if the PSP went below the fixed price for any half-hour, the supplier would pay the broker for the difference between the fixed price and the PSP multiplied by the hedged volume. In this way, the supplier was effectively paying a fixed price for its hedged volume. Obviously, if the hedged volume were an under-forecast, the supplier would be *exposed* for the difference between the forecast and actual demand. If the forecast were too high, then the supplier might be paying over the odds for the contract, although it could be possible to make a profit with some speculation. Generators could fix the price they were paid for a hedged volume of generation in exactly the same way.

Eventually, bespoke contracts became 'off-the-shelf' contracts with standard terms and conditions and standard contract types; for example, for base load or peaking. These were known as *Electricity Forward Agreements* (EFAs). These could be bought and sold on *Power Exchanges*. In this way, suppliers and generators could buy and sell in a *forward trading* market.

7.7.2.5 Electricity Market Reform (EMR)

The electricity pool system as initially created was felt to have played its part in the successful deregulation and opening up of the ESI. However, criticisms were made of the way in which pool prices could be manipulated by a few key generators bidding into the pool. In fact, it was considered that this 'rigging' of pool prices was keeping wholesale electricity prices artificially high. In addition, it was felt that generators and suppliers should be free to enter into *bilateral* agreements instead of also buying from the pool, albeit with a bilateral hedging contract sitting on top. Furthermore, it was seen as beneficial if third parties could enter the market to trade in physical electricity much as they had been doing in CFDs and EFAs. For this reason, the New Electricity Trading Arrangements (NETA) were introduced in March 2001, upgraded in 2005 to become a single electricity market for England Wales and Scotland known as the British Electricity Trading Transmission Arrangements (BETTA). Rising energy costs and concern with the operation of BETTA and in particular a lack of adequate investment in new plant prompted the UK government to consider a significant change in direction. Over a period of five years the EMR legislation was formulated, taking effect in 2014. There are four key elements to the EMR: fixed-price 15 year contracts for difference (these were outlined before in the context of renewable electricity generation but apply to all forms of generation); a capacity mechanism with capacity payments for availability to support investment in new generation capacity; a carbon floor price; and an emissions performance standard that in effect precludes new coal-fired generation (unless fitted with carbon capture). In many regards, the EMR represents a move away from a free and deregulated electricity market and gives OFGEM (the UK market regulator) and National Grid keys roles in managing the system.

7.7.2.6 Ancillary Services

In the UK, the Electricity System Operator, formed in 2019 within National Grid, is responsible for ensuring the secure operation of Britain's power system. They procure all the required balancing services, including those known as ancillary services, from commercial suppliers. These include: frequency response services; reserve services; reactive power services. Also covered is demand side response and 'black start' to support restarting a power system following a catastrophic collapse.

7.7.2.7 Marketing Green Electricity

The steady growth of renewable power generation has in part been supported by a growing market for so-called green electricity. The opening up of electricity markets created the opportunity for new and existing suppliers to offer green tariffs to consumers. Green electricity is marketed as a specific product to consumers who want to support renewable energy and a slightly higher rate may be charged. The income from such tariffs is either paid directly to the renewable energy generator or used to support new renewable energy generation. The take-up of these tariffs has varied from country to country. In the UK for instance, some studies indicate that over 50% of domestic electricity customers would be interested in a green electricity tariff, even if this means paying a little more than the lowest prices to ensure their electricity comes from renewable sources. The actual market is far smaller with some studies indicating that less than 1% of households in the UK has switched to a green tariff. More recently green tariffs have become cheaper, often among the cheaper tariffs available. In this context the market can be expected to grow rapidly.

References

1 Renewable Power Generation Costs in 2017: An Overview; International Renewable Energy Agency (IRENA) report, 2017.
2 Externalities of energy (ExternE) European Commission, 1995–99. EUR 16520-25 EN, also, External costs. EC Summary document on the above, published 2003. EUR 20198.
3 Edenhofer, O., Pichs-Madruga, R., Sokona, Y. et al. (eds.) (2011). *IPCC Special Report on Renewable Energy Sources and Climate Change Mitigation*. Cambridge, United Kingdom/New York, NY, USA: Cambridge University Press.
4 External costs – research results on socio-environmental damages due to electricity and transport. European Commission report EUR 20198, 2003.
5 REVALUE-The value of renewable electricity. European Commission project JOR3-CT96-0093.
6 "HM Government report Scotland analysis: Energy", presented to Parliament by the Secretary of State for the Department of Energy and Climate Change, April 2014.
7 (2011). *Scottish Generation Scenarios and Power Flows: An Analysis*. Sinclair Knight Merz.
8 Heptonstall P., Gross R. and Steiner F. "The costs and impacts of intermittency – 2016 update". UK Energy Research Centre report, February 2017.
9 Madaeni, S.H., Sioshansi, R., and Denholm, P. (2013). Estimating the capacity value of concentrating solar power plants with thermal energy storage: a case study of the southwestern United States. *IEEE Trans. Power Syst.* 28 (2): 1205–1215.
10 Dale, L., Milborrow, D., Slark, R., and Strbac, G. (2003). A shift to wind is not unfeasible. *Power UK* (109): 17–25.
11 Couture T.D., Cory K., Kreycik C. and Williams E. A Policymakers Guide to Feed-in Tariff Policy Design, Technical Report NREL/TP-6A2–44849, July 2010.

8

The Future – Towards a Sustainable Electricity Supply System

8.1 Introduction

There are good moral, economic and social reasons for dealing with world poverty. Extreme poverty is currently defined by the World Bank as living on less than $1.9[1] per day; at present approximately 10% of the world's population are exposed to this dire situation, most of these in central Africa. Inevitably, pulling people out such poverty will entail increased energy consumption to supply the goods and services required as part of this transition. At present the poorest individuals consume annually an amount of primary energy[2] between 3000 and 5000 kWh, in contrast, for example, to consumption in north America, which is running at over 80 000 kWh per year (and inhabitants of the Gulf oil states significantly exceed this value). Per capita electricity consumption tends to reflect this disparity with north Americans consuming over 13 000 kWhe (kWh of electricity) per year (a rate of 1.5 kW) whereas individuals in developing countries often average less than 500 kWhe (about 57 W), sometimes much less. The very low average figures for electricity consumption in the underdeveloped countries reflect the fact that, despite major progress in this regard over the last decade (especially in China), 1.2 billion people (16% of the global population) still have no access to electricity at all according to 2016 data from World Energy Outlook. It is simply not possible, or desirable, to have everyone on the planet consuming energy at the present north American rate, or even the lower European rate. But a massive increase in energy consumed by the poorest is required, and electricity will be a key part of this.

A large part of the problem is the increasing world population, and the continued perceived need for economic growth in the developed world. Somehow these issues must be addressed if human activity on the planet is to be sustainable. This, however, is not the subject of this book.

It is instructive to consider a universal electrification scenario that provides a modest minimum amount of electricity to all, say 1000 kWh per person per year. To achieve this by 2050 would require something like 10 000 GW of new generating capacity globally, more than twice today's level. Is it feasible to achieve this using traditional fuels or nuclear power? In the context of climate change, the first option would be an environmental disaster and would also substantially shorten the period over which the limited

---

1 Technically this is defined in terms of 'international dollars' where correction is made for local purchasing power.
2 This is the primary energy in effect consumed by an average person in the country concerned, i.e. national total primary energy divided by population.

Renewable Energy in Power Systems, Second Edition. David Infield and Leon Freris.
© 2020 John Wiley & Sons Ltd. Published 2020 by John Wiley & Sons Ltd.
Companion website: www.wiley.com/go/infield-renewable-energy

oil and gas reserves would be available. The second option could be environmentally benign (depending on the true environmental impact of the technology) but would require the installation of numerous nuclear power stations each week between now and 2050. Even if such a build rate were possible, the associated problems of security and nuclear waste disposal would be unmanageable. The fact that uranium reserves are finite will also impose a severe limitation on such a scenario.

The only convincing alternative is to continue to increase the exploitation of renewable energy (RE) sources, leading eventually to a completely sustainable electricity supply system.

Recall that renewable resources have a range of advantages; they are:

- not subject to depletion and will exist as along as the sun continues to shine;
- indigenous, hence reducing dependency on imported fuel from other countries with sometimes unstable regimes;
- virtually nonpolluting, with some small emissions produced during manufacture and end-of-life disposal;
- diverse and complementary in their time variability;
- potentially small scale (in particular photovoltaic, PV) and so can often be located near to demand, reducing transmission and distribution losses;
- robust in system terms because of the very large numbers of individual generators used and the statistical robustness of such a collection compared to centralised generation;
- particularly suited to the needs of developing countries, where systems based on renewable energy technology can be simpler to manufacture than traditional fossil fuelled or nuclear powered stations.

However, renewable energy sources involve still relatively new and somewhat underdeveloped technologies and most of the companies involved lack the economic and political influence of the leading conventional energy companies (ExxonMobil, Chevron, Shell, BP etc.). There are important exceptions, most notably Dong Energy (now Ørsted A/S) who have transformed themselves over the last decade from being the main Danish supplier of conventional electricity, to being primarily an offshore wind farm developer. The key disadvantages of renewable sources are that:

- the cost of electricity from some, but not all, renewable sources is at present higher than that from conventional energy sources. This comparison, however, depends importantly on the inclusion or exclusion of external costs. Moreover, costs for the more widely used technologies of PV and wind have been falling sharply (as outlined in Chapter 2);
- except for biomass based generation, they are not dispatchable and increase the challenge of load matching;
- their distributed nature may require restructuring of the electricity supply infrastructure (although this can also be seen as an advantage in cases where generation is near to loads).

These last two topics should be of particular interest to readers of this book who are now in a position to understand the key technical features required of future electricity supply systems. They will be explored in more detail in the remainder of this chapter alongside projections for the expected evolution of the individual renewable energy technologies. The positive news is that the world is now adding more renewable energy capacity each year than that powered by coal, natural gas and oil combined, so there is clearly progress in the desired direction. Nevertheless, the challenge of creating a truly sustainable electricity supply system should not be underestimated.

8.2 The Future of Wind Power

8.2.1 Large Wind Turbines

Recent progress in wind turbine technology has been impressive. At the time of writing, 8-MW wind turbines are commercially available. Twenty years ago, the largest wind turbine on the market was rated at around 800 kW with most manufacturers offering 600 kW units. At present, most wind turbines are fitted with rotors of 60–140 m diameter compared to the 35–45 m of the mid-1990s. The highest rated wind turbine currently commercially available is the Vestas V164-9.5 MW.

Due to economies of scale, larger machines tend to generate electricity at a lower cost per kWh, particularly in offshore applications. This progress is reflected in the fact that in 2017, for the first time, major offshore wind farms are being constructed without government subsidy. These are: the Dong Energy (Ørsted A/S) 240-MW OWP West and the 240-MW Borkum Riffgrund West 2 wind farms, and the EnBW 900-MW He Dreiht project.

This increase in size has been evolutionary in nature with successive but incremental scaling up of proven designs, allowing manufacturers to progressively improve their understanding of and confidence in the design process and key issues such as fatigue. Is this the best recipe for future expansion though? Might step change innovations provide a faster path to lower cost larger machines? And is there a limit to the size of wind turbines imposed by material strength limitations? Since the emergence of the first commercial wind turbine in the mid-1970s, experts have been predicting definite limits to turbine size. All these limits have now been exceeded, pointing up the hazardous nature of such predictions.

Figure 8.1 shows an artist's impression of a radical design for a 10 MW wind turbine proposed by a consortium led by Dutch rotor blade maker Polymarin Composites some years back. This turbine, known as ICORASS, which stands for integral composite off-shore rotor active speed stall-control, has a 160-m two-blade rotor and was intended for offshore applications where foundation and installation costs do not scale linearly with turbine power, providing cost savings for larger unit sizes. The design was intended to be highly robust so as to provide exceptional reliability and availability as required for offshore application where access and maintenance is difficult. Key features of this particular design are the integration of a direct drive alternator into the hub, a fixed blade angle with stall power control and the use of a permanent magnet alternator for speed reduction (braking) above rated wind speed to ensure fixed power delivery, and last a downwind free yawing rotor to remove the need for an active yaw drive system. Although this concept was not commercially developed, it resembles in many regards (including the low-cost lattice tower) the 6-MW turbine presently offered by 2B Energy for offshore applications.

Another, perhaps more innovative design, is the multi-rotor. In this approach, a number of smaller rotors are attached to a single support structure. Figure 8.2 shows

Figure 8.1 Artist's impression of the ICORASS 10 MW wind turbine. Source: Reproduced from Polymarin.

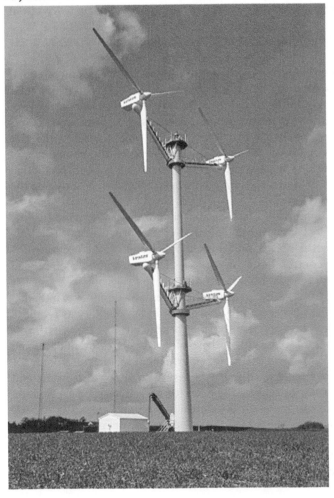

Figure 8.2 Vestas multi-rotor prototype.

a 900 kW rated prototype under test by Vestas in 2016; it is made up from four V29-225 kW nacelles supported by a tubular steel structure that facilitates yawing. The motivation for this approach is the 'square-cube' law that states that whereas the power from a turbine scales with the swept area and thus the square of the diameter, the mass (and thus roughly the cost) scales with the volume that scales with diameter cubed. Thus, all other factors being equal, turbines become more expensive per unit output as they grow in size. Despite the limitations of this simplified scaling, the argument has some merit. A detailed account of the multi-rotor concept can be found in Reference [1].

Airborne wind systems have also been considered and some prototypes tested. Figure 8.3 shows a Buoyant Airborne Turbine (BAT) developed by Altaeros Energies (an MIT startup company). It is intended to operate at a height of 1000 ft, held in place by a helium filled balloon like structure.

There are other approaches to airborne wind power, including tethered kites that are controlled to drive generators on the ground. In both approaches (airborne generation or ground based), the key challenge is to efficiently generate and deliver the electricity.

It is too early to say whether such advanced design concepts are likely to be commercially successful, but it is clear that future turbines will be significantly larger than today's, especially where offshore application is concerned, and that wider use will be made of advanced composites. Alongside innovations in construction, increased use will be made of sophisticated condition monitoring and preventative maintenance systems.

Figure 8.3 Prototype Buoyant Airborne Turbine developed by Altaeros Energies.

8.2.2 Offshore Wind Farm Development

An offshore installation has advantages over and above the reduction in turbine costs from the resulting economies of scale. The marine environment is windier, resulting in an increased energy yield, and the winds are more persistent and less turbulent making system-wide integration easier. It has also been identified that inter-annual variations in wind speed are less offshore than onshore, making annual energy yield more dependable. Additionally, the transportation of exceptionally large or heavy items is generally easier over water than over land where roads may need widening of reinforcement. Importantly, planning approval may be easier to obtain for offshore schemes due to their reduced visual and noise impact. Offshore developments are often less objected to by the public since very large wind turbines that might cause an unacceptable visual intrusion onshore are often found to be acceptable if sited sufficiently away from the shore. Such advantages will become increasingly important as the viable onshore sites are progressively used up. At present, offshore wind energy is more expensive than onshore, primarily due to significantly higher foundation, installation and electrical connection costs. Complex and costly underwater cable systems are required to link the turbines together and to make the connection to shore. In some cases, an offshore substation may be required. However, as more offshore plant is built, the technology will advance to deal with unanticipated problems and to reduce transportation and installation costs. For example, while lightning tends to strike onshore wind turbines at the blade end, in offshore applications it was observed that strikes are more frequent at the centre of the blade. Blades are now redesigned to cope with such occurrences. To speed up installation, dedicated vessels are being developed and built to deliver and install the towers nacelles and blades.

One of the major problems is the accessibility of turbines for maintenance. This is only feasible by sea during relatively calm conditions, so the larger offshore wind turbines are sometimes fitted with platforms on which personnel can be lowered from a helicopter. These procedures are costly so designers are under pressure to develop turbines with exceptional reliability and long maintenance intervals. As already mentioned, sophisticated condition monitoring systems continuously assessing the turbine performance and fatigue damage will become increasingly common. Such systems were too expensive for onshore turbines of 1–2 MW, but for the present multi-megawatt units the systems represent a much smaller fraction of the turbine capital cost; moreover, the costs of instrumentation and data processing are falling making such approaches more cost effective.

Offshore foundations are another area where considerable progress has been made and future developments are expected. Monopiles have grown in size, now up to 8 m in diameter, allowing larger turbines to be supported, and in greater water depths, up to 40 m. They now dominate the market with over 95% of offshore installations using this technology. The remainder, in greater water depths, use jacket structures. Beyond about 50 m water depth, bottom mounted approaches become excessively costly and attention has moved to floating support structures. Prototypes have been tested, but to date no completed commercial wind farms have deployed floating turbines. A number of different technologies are available.

The first full-scale demonstration of floating wind, known as Hywind (Figure 8.4), was installed by Statoil in 2009 near to the Norwegian coast, comprising a 2-MW

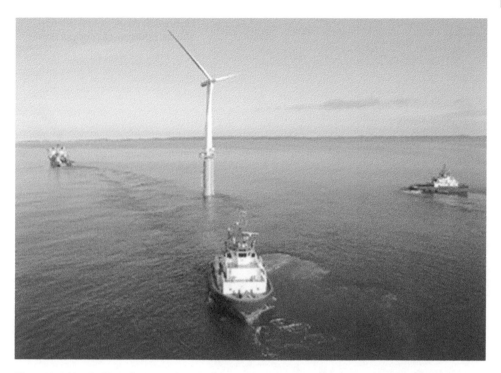

Figure 8.4 Hywind installation 12 km south east of Karmøy, Norway. Source: Credit: http://www .siemens.com/press.

Siemens turbine on a spar buoy support structure, in 210 m of water. The installation was operated successfully for many years providing valuable data. This particular support structure was designed to be very rigid, allowing little movement of the turbine with waves or wind gusts, and little adverse impact on turbine structural loads was identified. Of course, this rigidity has an engineering and thus economic price. The technology has now been scaled up, resulting in an 80-m spar suitable for mounting 6-MW Siemens turbines (type SWT-6.0-154). The technology was used to form the first semi-commercial wind farm utilising floating turbines, known as Highwind Scotland Pilot Park. This is located 25 km of the Scottish coast at Peterhead in 95–120 m of water. Five turbines will give the wind farm a 30 MW rated power. Installation started in May 2017, and the wind farm is now operational.

The second demonstration project of scale made use of a potentially lower cost support structure comprising three linked squat steel cylinders with heave plates to limit movement in rough seas. This was the 2-MW WindFloat project installed 5 km off the coast of Aguçadoura, Portugal in 2011 (Figure 8.5). The support structure is kept in place using conventional moorings to limit costs, so it does experience considerably larger displacements than the spar buoy of Hywind. This helps to limit the loads on the support structure, but does lead to larger aerodynamic loads on the turbine. The prototype was operated successfully over five years, and on this basis the design is being scaled up to 8 MW for future commercial deployment by the company Principle Power. A useful up-to-date review of floating wind turbine foundation structures can be found in Reference [2].

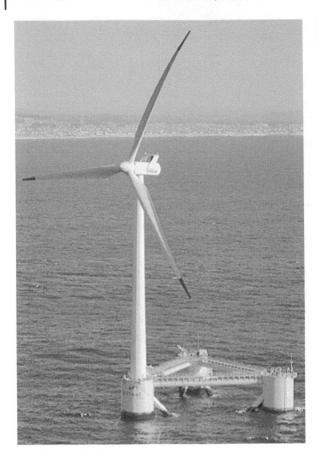

Figure 8.5 WindFloat installation. Source: Courtesy of Principle Power.

8.2.2.1 Electrical Integration

The growth of significant offshore wind capacity has raised grid integration issues to a new prominence. Increasingly large amounts of electricity will be feeding into national networks at points not specifically designed for such infeeds. Consequently, significant network reinforcement will often be required. In addition, these time varying infeeds must be integrated into grid management systems not previously required to cope with such non-dispatchable sources. The costs of expected system adaptation will be significant and new market mechanisms may well be needed to ensure that the required infrastructural investments are made in a timely manner.

Today's onshore wind turbines typically generate at around 700 V, which is commonly stepped up to ~35 kV by a transformer in the nacelle or the tower base. The first offshore schemes have tended to use a similar arrangement, although it is anticipated that as wind farms become larger with longer distances between turbines and increased distance to shore, the voltage will have to increase.

In the Danish 40 MW Middelgrunden wind farm a voltage level of 30 kV was chosen for the collection of power between the 20 turbines and the short 3-km connection to shore. In contrast, for the 160-MW Horns Rev wind farm, the collection voltage is 36 kV but the voltage is then raised to 150 kV at an offshore transformer substation before transmission over 15 km to shore. It is now increasingly common for offshore substations to be used. In future, large offshore schemes with AC transmission may severely limit the distance over which electrical power may be transmitted by undersea cable.

8.2.2.2   DC Transmission for Wind

Voltage source converter (VSC) based high voltage direct current (HVDC) technology has advanced steadily in recent years [3, 4]. Manufacturers now offer such links rated at 1 GW or more and operating at ±450 kV on the DC side. This substantially extends the offshore range and size of wind farms that can be potentially exploited. Whether such schemes will be widely used in the future depends on the shortage of near-shore sites and on the technical success and reliability of the present offshore schemes.

Despite the long-term expectations, early experience with HVDC for offshore wind in Germany has proved challenging with both cost and time overruns being commonplace. The key technical issues encountered in the early HVDC systems are mostly related to system operation, control and protection [5].

The first such connection (BorWin1) was between the BARD 400 MW Offshore 1 wind farm and the network connection point at Diele, 200 km distant. The link has been operating since being commissioned in 2010, but not without technical difficulties. It uses the ABB HVDC Light system (self-commutated converter based on Insulated Gate Bipolar Transistors, IGBTs). It has been estimated that the use of HVDC for this project has reduced transmission losses by about 25% compared to an equivalent AC link. Since then a number of further offshore wind farms in Germany have been connected using HVDC. The general perception of adverse technical issues and cost overruns has to date prevented the application of HVDC links to connect offshore wind farms in other countries.

Figure 8.6 shows the DolWin beta offshore converter station, located 45 kms from the German coast. This 320 kV converter is rated at 916 MW and weighs around 23 000 tons.

8.2.2.3   Innovative Collector Systems

The possibility of HVDC being used to connect offshore wind farms to the onshore power network highlights potentially unnecessary power-electronic conversion stages. The turbines generate AC, this is converted to DC as part of the conventional variable speed drive, then to AC for the wind farm collection system, then back to DC at the HVDC offshore converter station and finally back to AC at the onshore HVDC converter station. Some researchers have explored alternative arrangements that reduce the amount of power-electronic conversion. One possibility is a DC offshore collector system, such as that shown in Figure 8.7.

An alternative is to cluster a number of nearby wind turbines together using AC connected induction machines (at around 700 V) and to use a single variable speed drive converter. The individual turbines would be to an extent locked together in speed, although the slip of the induction machines would allow some small variation. Figure 8.8 shows the proposed arrangement. Analysis suggests that it would be feasible to cluster together around five wind turbines in this way.

8.2.2.4   A Proposed European DC Supergrid

VSC based HVDC technology can be adapted to operate with several terminals on the DC side so that power can be injected or extracted at will at each terminal. This technology is known as multi-terminal HVDC; a useful review of DC-DC converters for

Figure 8.6 DolWin beta offshore wind converter station. Source: Courtesy of ABB.

Figure 8.7 Schematic of DC collection system taken from [6]. Source: Courtesy of Y Lian and S Finney.

such application can be found in Reference [8]. This proposed mode of operation has encouraged feasibility studies for a European DC supergrid. Because of the concentration of the wind power resource in the Atlantic, the North Sea and the Baltic, there has been a suggestion that such a grid would be highly beneficial (see, for example, Reference [9]). It would range from Scandinavia down to France and would link Germany, the

Figure 8.8 Variable speed AC cluster concept taken from [7]. Source: Courtesy of D W Elliot, S J Finney and C Booth and IET.

Figure 8.9 Concept map of the proposed European DC supergrid. Source: Reproduced with permission of Airtricity.

UK, Ireland and Spain on the way, as shown in Figure 8.9. The cost of such an enterprise would be substantial but it has been calculated that it would be more cost effective than the upgrading of the existing onshore networks necessary to absorb the large available offshore potential.

Another advantage would be the considerable aggregation of the resource over a very large geographical area, resulting in a much firmer resource in spite of local variations. Additionally, the system would have the ability to provide energy to the country with the greatest need at specific times and the possibility of using the grid as an interconnector between national markets to enhance energy trading. This is particularly attractive because of the time difference and therefore the time shifts in peak demand among the linked counties. It should be noted that present HVDC systems are usually point to point, therefore considerable technology development would be required to deliver a multi-terminal HVDC network of the sort proposed, as discussed in Reference [8].

8.2.2.5 Smarter Wind Farms

Due to the economies of scale, future offshore wind farms are likely to be an order of magnitude larger than the present onshore schemes, and onshore wind farms are also increasing in size, resulting in the rating of such wind farms becoming comparable to that of traditional generating plant. As a consequence, they will be required to perform functions not demanded from present onshore wind farms.[3] Such functions will include:

- control of active and reactive power;
- system frequency control;
- local grid voltage control;
- operation under abnormal conditions such as grid faults;
- contribution to network stability.

These requirements will demand significantly improved sophistication of wind turbine and wind farm control functions. Since the majority of the current multi-megawatt wind turbines operate at variable speed using either doubly fed asynchronous generators or synchronous (mainly permanent magnet) or asynchronous machines connected via a full converter, they can be controlled to meet these demands. Chapter 4 explained how power electronics can control real and reactive power and thus the local voltage, and in Chapter 6 the use of power electronics to achieve fault ride-through is outlined. Chapter 3 outlined how modified control of the turbine could provide frequency response services, in particular, emulated inertial response and droop. Both voltage control and frequency control contribute to network stability. Power system damping can be provided by appropriate dynamic control of the turbines, see for example References [10, 11], and this can include the damping of inter-area modes [12].[4]

In recent years, wind researchers have been considering how wind farm control might be further improved by considering the performance of the wind farm as a whole, not just the individual turbines. Reference [13] is a recent book that deals specifically with this transition, and in particular how wind farms can in theory achieve greater power production than by maximising the output of individual turbines. Wind farm modelling suggests that overall wind farm power generation can be increased by not operating all

---

3 Indeed, a number of these requirements are incorporated into recent Grid Codes, for example those of E.ON Netz, Germany (Grid Code High and Extra High Voltage, E.ON Netz GmbH Bayreuth, August 2003), National Grid, UK (The Grid Code, Issue 5, Revision 21, January 2017) and EirGrid, Ireland (Grid Code V6.0, July 2015).

4 Inter-area modes occur when say two groups of conventional generators separated by a transmission line interact dynamically to create an oscillation (involving the phase angle of the generators and also the currents in the line).

turbines at their maximum power operating point. By reducing the power and thrust of an upstream turbine, more wind can reach the downstream turbine, and the suggestion is that this may increase the overall power output. Such a strategy, though, requires testing on real wind farms before being adopted commercially. Chinese wind turbine manufacturer Envision claim that their smart wind farm control can increase wind farm performance by 20%, although exactly what this means is somewhat open. Certainly, researchers have been most active in exploring the coordination of individual wind turbine control at wind farm level so as to best deliver real and reactive power at the point of common coupling. A study by ECN, [14], summarised the key attributes of a wind farm controller as reflecting the limitations for:

- current and voltage rating of the power-electronic converter and other electrical devices;
- the aerodynamic torque (and rotor speed);
- the grid characteristics in combination with the wind turbine operational range and the grid code requirements.

Wind farm active power control should:

- maximise the total wind farm power output, for example using active wake control;
- regulate the wind farm power output according to an external set-point, which usually includes power (ramp) limitations, maintaining a certain balancing reserve and so on;
- mitigate wind turbine loads.

And wind farm reactive power control should:

- keep the voltages in the wind farm collection grid and in the transmission grid within the normal operating ranges;
- minimise transmission losses, for example through contributing to the reactive power compensation;
- control the reactive power at the grid connection point according the grid operator set-point.

It should be noted that the development in practice of effective wind farm controllers depends critically on the interaction between turbines through their wakes, since control of a turbine will directly impact on the wind flow to downwind turbines. This is also an area of active research as it is recognised that the existing simplified wake models widely used by the industry are not sufficiently accurate. One idea being explored is to use yaw control to deflect wakes away from downwind turbines, thus increasing wind farm output, see for example Reference [15]. This approach has been supported by some modelling studies, but relies on very accurate prediction of wakes and how they move about under changing wind conditions (including wind direction changes). Wind turbines operating in the wakes of other turbines also experience added turbulent loading; this is especially marked if a turbine is only partially immersed in a wake, since in that case each blade is moving from free wind to wake flow once per revolution, substantially increasing fatigue damage. It is hoped that a smart wind farm control will be developed in future that takes all of these issues into account, potentially reducing the cost of wind energy appreciably. As mentioned previously, the Chinese company Envision claim to be already doing this, by combining 'smart turbine control technology with cloud computing, successfully upgrading the smart turbines groups into smart wind farms'. Elsewhere, there are research projects specifically aimed at developing such smart wind farms, for example, Smart Wind Farms – simulation and optimisation of wind farms being undertaken at Fraunhofer IWES in Germany and also smart wind farm control (SWIFT) in Belgium.

8.2.3 Building Integrated Wind Turbines

There has been considerable interest, particularly in Northern Europe, in the possibility of fitting small wind turbines on the roof tops of individual houses. Enthusiasts for this approach claim that small wind turbines sited within the urban environment have the potential to make a significant contribution to the reduction of CO_2 levels. A typical wind turbine of this type is illustrated in Figure 8.10.

The difficulties facing the designers and manufacturers of such machines are substantial. The dis-economies of downsizing are such that the installation cost per kW of a small wind turbine is considerably more than for a MW size machine. Moreover, the annual average wind speed in a typical urban environment is perhaps one half of that usually experienced at a good wind farm site. The cubic power law therefore indicates that the expected energy yield per kW capacity from the small wind turbine in these circumstances would be approximately one-eighth of that from its large-scale counterpart, and this ignores the substantial advantage of the higher wind speeds at the hub height of the large machine. Additionally, in urban locations wind turbulence is considerable due to the presence of buildings and trees and this can affect energy capture and the life of the turbine. On the positive side, small wind turbines if manufactured in, say, hundreds of thousands, can gain enormously from mass production economies of scale and present costs may be more than halved. Government subsidies may accelerate their adoption and their progress down the learning curve.

In general, however, because of long current payback periods, the market for such machines is limited with the unfortunate side effect that the low costs that may result

Figure 8.10 Small building integrated wind turbine. Source: Reproduced with permission of Windsave.

from mass production are unachievable at present. There are in addition concerns about potential noise and vibration problems, as well as the challenge of obtaining planning permission,[5] especially if neighbors object to the installation, and as yet untested arrangements for selling any excess energy fed to the grid. However, there will be some urban areas on hilltops and on coastal locations where the wind speed is high enough, when taken together with escalating electricity costs, to justify the investment in this technology. Such domestic wind turbine applications are reviewed in Reference [16].

Small wind turbines are already economically viable in remote locations not serviced by the grid and for the provision of modest amounts of electricity to caravans and sailing boats. This is an already established market, but one for which the costs of alternative electricity supply is substantially higher than that supplied from the grid.

Enterprising architects have produced futuristic designs that incorporate wind turbines within the building structure, frequently with a ducting effect to enhance the energy capture. One such design is shown in Figure 8.11, where two 43-floor office towers are connected by three turbine bridges. Each turbine is 29 m in diameter and has a

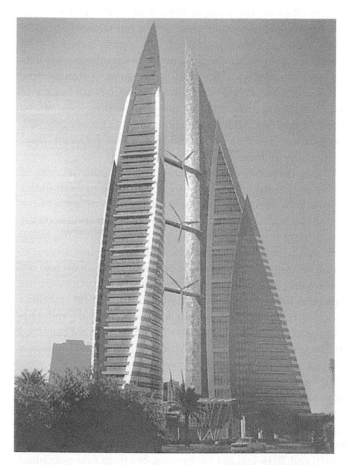

Figure 8.11 The Bahrain World Trade Centre. Source: Reproduced by permission of the Bahrain World Trade Centre image © Khalid Al Muharrqi based on a design by Atkins.

5 Interestingly, the UK Government has been considering legislation that could remove the requirement for planning consent for suitable small turbines.

rating of 225 kW. The turbines were intended to supplement the energy requirements of this prestigious Bahrain World Trade Centre.

8.3 The Future of Solar Power

As described in Chapter 2, solar technology is developing fast with impressive reductions in costs in recent years. In parallel, the scale of the centralised installations has increased significantly with the largest plants now being well over 500 MW rated and soon to approach 1 GW, thus starting to make a significant contribution to bulk electricity generation. Future developments are discussed next. A good review of the rapidly falling PV costs achieved in recent years can be found in Reference [17].

8.3.1 PV Technology Development

The existing PV market is dominated by silicon wafer cells. Commercial modules based on this technology are up to 24% efficient, the present record of 24.1% held by Sun-Power was established in 2016. Monocrystalline cells are efficient but somewhat costly to manufacture and have a high embodied energy content. The energy payback is in years, with the exact figure dependent on the local radiation resource, and contrasts with wind energy where the energy payback period is typically measured in months (see, for example, Reference [18]). Polycrystalline cells are cheaper to manufacture but are of lower efficiency. Nevertheless, their module efficiency is now approaching 20%, not so long ago the benchmark for monocrystalline modules. Silicon accounts for roughly 50% of the total manufacturing cost of these wafer-based modules, and thus for many years there has been interest in thin-film silicon technology.

Thin-film PV has the potential to deliver cheaper electricity but despite decades of development has yet to really deliver commercially. In 2015, according to Fraunhofer Institute for Solar Energy Systems, thin-film technologies accounted for only 8% of the total PV market. Amorphous silicon was the first thin film to become commercially available but is no longer the dominant thin film technology commercially. It has failed to grow market share as anticipated (indeed has lost market share in the last 10 years), in the main because of its poor efficiency, below 10% for modules. Intrinsic materials issues have limited its performance and it is not expected to be competitive in the longer term. Currently, commercial interest centres around thin-film nanocrystalline hybrid cells and the heterojunction cells based on copper indium diselenide and cadmium telluride. Both of the latter provide reasonably stable efficiencies and are expected to benefit from improved manufacturing techniques. Investments in these technologies includes manufacturing onto glass, and also metal and polymer foils.

Industry experts confidently expect that the new mass production technologies together with efficiency improvements will deliver cells capable of generating electricity competitively with conventional forms and nuclear power within the next 10 years. An up to date review of cell efficiency improvement and the future potential of PV can be found in Reference [19]. Figure 8.12, taken from this study, shows the increases in cell efficiency achieved between 2010 and 2016 (except for perovskites from first published data in 2013, and CdTe from 2011). Progress made before 2010 is indicated by the vertical lines. CdTe (cadmium telluride), QDs (quantum dot cells) and perovskites all

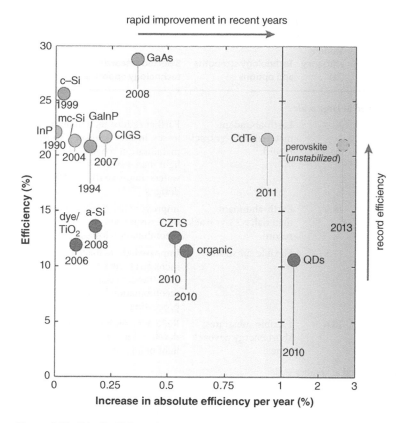

Figure 8.12 PV cell efficiency improvement.

show considerable potential. Improvements in laboratory cell efficiency, though, are not always a good guide to future commercial module efficiencies.

Table 8.1, also taken from this study, summaries both the cell and module efficiencies presently available, together with an indication of how the different technologies can benefit from future research and development.

8.3.1.1 Different Deployment Options

In the early years, the bulk of current grid-connected PV was located on domestic building roofs. Japan and Germany between them installed over a million such roof systems, primarily as a result of generous market support mechanisms. And similar programmes have followed in other countries. The first rooftop installations made use of traditional looking modules, although later there was an increasing use of so-called PV roof tiles. These can be laid more or less in the same way as traditional tiles and lend the roof a more traditional appearance. Alongside domestic roofs, commercial buildings also provide attractive locations for PV where flat roofs and facades lend themselves to PV installation. There are numerous examples of architecturally appealing PV facades to be found around the cities of Europe. There are thermal issues associated with such building integration of PV. If not carefully designed, the mounting systems can inhibit natural ventilation of the modules, leading to overheating with possible damage to the PV cells

Table 8.1 Key research and development opportunities for PV development

Material	Cell efficiency (%)	Module efficiency (%)	Technology strengths and options	Selected research/ technology opportunities
Mature technologies deployed at large scale				
Monocrystalline Si	25.6	22.4	Earth-abundant material; >25-yr track record	Further reduce recombination losses, in combination with new metallisation schemes; improve light management in thinner wafers; improve IBC and SHJ cell designs
Multicrystalline Si	21.3	18.5	Earth-abundant material; >25-yr track record	Improve wafer quality (minimise or passivate defects) to reduce recombination losses
CIGS	21.7	17.5	Flexible substrates	Improve light management; increase efficiency for large band gaps (tandem cells); reduce recombination losses, solution processing
CdTe	21.5	18.6	Flexible substrates; short energy payback time	Reduce recombination losses; develop thinner cell designs using light management
Emerging technologies deployed at smaller scale				
Dye-sensitised TiO$_2$	11.9	10.0	Tunable colours	Improve redox couple; reduce recombination losses; increase band gap; increase stability
Thin-film Si	11.4	12.2[a]	Flexible modules	Reduce recombination losses; improve light management
Organic	11.5	9.5	Flexible modules, semitransparent modules	Improve light management; increase band gap; increase stability; reduce recombination losses
Technology at the manufacturing level				
GaAs	28.8	24.1	Very high efficiency; flexible modules	Improve light management; develop IBC geometry; further develop thin-film multijunction cells by layer transfer
Technologies under development				
Perovskite	21.0	n.a.	Solution processing flexible modules	Reduce recombination losses; improve cell stability; avoid use of Pb; increase efficiency for high-band gap materials (tandem cells); develop Si/perovskite tandems
CZTS	12.6	n.a.	Flexible modules	Reduce recombination losses; improve light management
Quantum dots	9.9	n.a.	Solution processing; flexible modules	Reduce recombination losses; improve light management; avoid use of Pb

a) Microcrystalline Si/a-Si tandem geometry.
Source: Courtesy of Polman et al. [19].

Figure 8.13 Artist's impression of TESLA gigafactory.

in conditions of high radiation and ambient temperature. Fortunately, these issues are now fairly well understood.

In contrast to the various forms of building integrated PV, large-scale arrays of PV modules can be located on dedicated, usually rural, sites. Systems of up to hundreds of MWs are being installed. Battery manufacture TESLA are presently planning a 70 MW PV array to be integrated into their new 'gigafactory' in the Nevada desert (Figure 8.13 is an artist's impression of this futuristic facility). The USA leads in this development, alongside China. The sunnier parts of Europe have also recently seen significant installations. In 2009 the world's biggest solar power plant was in Walpolenz, Germany, covering approximately 110 ha ($1.1 \, \text{km}^2$) and rated at 40 MW. An indication of the development since then is that such a plant would be considered modest by today's standards. The largest plant in 2017 is the Longyangxia Dam Solar Park in China, covering $14 \, \text{km}^2$ (Figure 8.14), and rated at 850 MW; the total construction cost was 6 billion Chinese RMB. The PV park is located alongside a 1280-MW hydro plant and shares its network connection. It is expected to generate over 800 GWh per year, enough for around 200 000 households. These massive investments indicate the attraction of PV technology, and suggest the scale for future developments.

8.3.2 Solar Thermal Electric Systems

As outlined in Chapter 2, considerable effort has gone into the development of solar thermal electricity generating systems. These are only feasible in large-scale developments and as a result the present solar thermal installations are comparable in size to the larger PV power plants (in the range of 100–400 MW). The USA and Spain are the countries most committed to this technology. Presently, the largest installation is the

Figure 8.14 Satellite view of the 850 MW Longyangxia Dam Solar Park. Source: Credit: NASA Earth Observatory.

Ivanpah Solar Electric Generating Station (Figure 8.15), located in the Mojave Desert and rated at 392 MW. It comprises 173 500 heliostats over an area of 16 km^2 focused on three power towers. It was commissioned in 2014. It has been estimated that if an area of hot desert of about 250 × 250 km, less than 1% of the total areas of the world's deserts, were to be covered by such thermal plant enough energy would be produced to cover all present electricity needs.

8.4 The Future of Biofuels

Biomass has a special role to play in the sustainable electricity supply systems of the future due to the fact that it is easily stored. Electricity generation based on biofuels, supported by adequate fuel distribution and fuel stores local to the generation plant, is fully dispatchable and thus can be used to compensate for the variations in the non-dispatchable renewables and the system load. Whether there will be sufficient

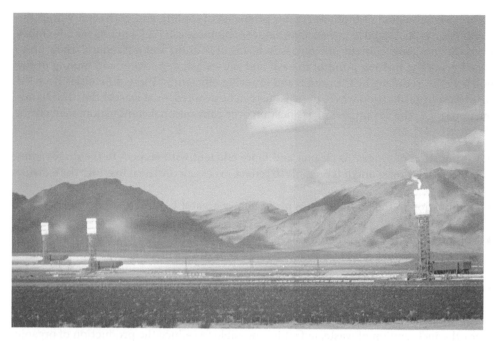

Figure 8.15 World's largest solar thermal power station, the 392 MW Ivanpah Solar Electric Generating Station in the Mojave Desert. Source: Image from Sbharris – Own work, CC BY-SA 3.0, https://commons .wikimedia.org/w/index.php?curid=31089494.

biofuels for this purpose will depend on the available land area for energy crop production in relation to the amount of balancing plant required.

Biomass resources in most countries are large; for example, in the UK is has been estimated that up to 20 million tonnes could be available per annum. It is important to distinguish between first-generation crops that have been developed for food (sugar beet, oil seed rape and wheat grain) that may be used for chemical conversion to biodiesel and bioethanol and second-generation lignocellulosic (biomass) crops that can be used as feedstock for heat, power and liquid fuels.

First-generation technologies have in general a poor carbon footprint and represent an 'intermediate step' towards second-generation lignocellulosic feedstock. Research is needed on the environmental impact of the wide-scale growing of biomass crops. At present there is limited understanding on how the different bioenergy chains compare in environmental impact.

There are also concerns that dedicating significant land area to bioenergy crops will put pressure on food production and result in unwanted increases in food costs. Impacts on food prices in the USA and China were reported some years ago, but these reflected the use of corn for the production of liquid fuels for transport, rather that fuels for electricity generation.

As reviewed in Chapter 2, biomass combustion for electricity generation can be undertaken in three distinct ways: direct combustion (which includes co-firing with coal); gasification; and pyrolysis. The latter two are indirect in that gaseous and liquid fuels respectively are created that in turn can be used as fuels for electricity generation.

Direct combustion of biofuels creates problems of fouling and corrosion within the combustor, exacerbated by the alkaline nature of biomass ash [20]. Fouling mainly comes from the inorganic content of biomass in terms of sodium and potassium. Their presence lowers the melting point of the ash and causes ash deposition and fouling of the boiler tubes, reducing heat transfer, and lowering efficiency. These difficulties have led to interest in gasification where these issues can be avoided, or at least at the initial combustion stage. The advantages of gasification over conventional combustion, taken from Reference [21], are:

- Combustion of syngas[6] is cleaner and more efficient with respect to the direct combustion of biomass and it is possible to reduce excess air and heat loss into the exhaust gases.
- Combustion in the homogeneous phase allows greater continuity of the process and easier control.
- Volumetric flow rate of the produced fuel gas is much reduced compared to that of the effluent gas of a direct combustion process, resulting in lower costs of the cleaning process.
- Even if the syngas produced is burned on site, there is the possibility of removal of certain impurities after the gasification stage and before the syngas combustion.
- Allows more restrictive environmental limits than at the 2000/76/EC European law for the incineration of waste, in fact it is possible to reduce the production of organic micropollutants such as dioxins and furans as well as NOx and SOx.

Gasifiers have been commercially available, but are not without their own challenges. Gasification for power generation provides a syngas that is most commonly then used to fuel a gas turbine. There are a number of different gasifer (or reactor) technologies that have been developed over the last decades. The main reactors used for gasification are [21]:

- Entrained flow reactor;
- Fixed bed;
- Fluidised bed;
- Rotary kiln reactor;
- Plasma reactor.

Gasification is also of interest in that it can make use of waste as well as biomass as the feedstock. A considerable number of gasifiers are in operation in Europe, across a wide range of sizes from hundreds of kWs to almost 300 MW. The most common biomass inputs are wood chips. They tend to operate as Combined Heat and Power (CHP), providing both heat (locally) and electricity.

A continuing challenge is that the syngas quality is not especially compatible with reliable combustion in a gas turbine, even after cleaning, due to contamination with pollutants such as tar, ammonia and sulfidic/chloridic acids. More research is ongoing to deal with these issues, and thus make biomass electricity generation more attractive in the future.

---

6 Syngas is a fuel gas mixture comprising mainly hydrogen and carbon monoxide, produced normally by gasification.

8.5   Geothermal Power

Geothermal power has not been dealt with in this book so far. This is because, although in effect a renewable energy resource, it has the characteristics of conventional base load steam plant. The resource is concentrated in particular geographic regions. Iceland is well known for its use of geothermal energy, both for heating and electricity generation, but it is also widely used in the USA, and also in the Philippines, Kenya and New Zealand. In 2015, over 12.5 GW of geothermal capacity was installed worldwide, and this is expected to grow significantly over the next decade.

8.6   The Future of Hydro and Marine Power

Although small scale hydro installations can make a useful local contribution to electricity generation there are a limited number of suitable sites and thus the overall impact on electricity system operation is limited. This is unlikely to change with time. Larger scale hydro is a different matter, and in 2016 accounted for 18.6% of worldwide electricity generation. In countries such as Switzerland and Norway with a large proportion of hydro power, the power systems are already designed around these resources and their characteristics. Some parts of the world have yet to develop their hydro potential fully and market analysts expect major investments over the next decade, especially in China, Brazil and India. With the installation of this new capacity, attention will also have to be paid to the creation of suitable system infrastructure.

Tidal power, as summarised in Chapter 2, has a large contribution to make in geographical areas that experience a large tidal range. To date, this resource has barely been touched. Since the technology for barrage construction and turbine construction is already available, it can be concluded that its unfavourable economics, and also environmental impact concerns, that presently inhibits the use of this technology. It is known that large tidal projects could take decades to pay back their installation costs. Even though the civil works would last well beyond such timescales, nearer term market pressures make these projects impossible to finance. Nevertheless, there will come a time when fossil fuel shortages and rising energy costs will make the construction of large barrage schemes attractive despite the adverse local environmental impact on wading birds and other wildlife. Indeed, the recent concerns for security of supply that are giving hope to the nuclear industry should be prompting a re-examination of this technology, which poses investment challenges that are not dissimilar. Both nuclear and hydro involve massive upfront costs and the need to be paid back with profit over their considerable lifetimes. As with nuclear, states will need to decide how to facilitate the construction of large tidal barrages as they have an important role to play in the long-term future electricity supply.

It may be that concerns about the ecological side-effects will result in alternatives to the straightforward barrage schemes. One such alternative has been suggested by Friends of the Earth in which tidal lagoons are created. These are claimed to be capable of producing even more energy than barrage schemes as well as being less environmentally intrusive. Figure 8.16 shows an artist's impression of a tidal lagoon for Swansea Bay.

The newer technology of tidal stream, introduced in Chapter 2, suites different sites and is generally seen as promising despite its present lack of development. It is widely

Figure 8.16 Artists impression of Swansea Bay tidal lagoon. Source: Taken from www.walesonline.co .uk/news/wales-news/head-swansea-bay-tidal-lagoon-13322009.

anticipated that reliable and cost-effective technology will be developed. This will make a useful contribution, but only in very specific geographical areas where suitable stream velocities exist. The Pentland Firth, just north of the north coast of the Scottish mainland, has an outstanding tidal stream resource. The first commercial tidal stream power station known as the MeyGen project, rated at 6 MW and comprising four AR1500 turbines, was commissioned in March 2017. A further phase of MeyGen is planned.

Wave power is perhaps the least developed renewable energy source, although the resource, especially around the UK coastline, is huge. Since 2000, a number of large-scale wave power prototypes have been demonstrated around the world, but the technology is still up to 20 years behind wind energy. In the UK, only one wave energy device (Pelamis, see Chapter 2) has been demonstrated at near full scale in the open sea, but ultimately this proved unable to survive commercially. Designing devices to survive in the open seas is challenging but enthusiasts for the technology are still projecting significant electricity generation from this source in the future.

8.7 The Shape of Future Networks

Some national power systems are already generating most of their electricity from renewable energy sources, for example in Norway with 98% of supply from hydro, and Kenya with over 50% from geothermal. However, both hydro and geothermal power generation are easy to accommodate in power systems being based around large synchronous machines. As this book will have made clear, the challenge lies in accommodating high penetrations of time variable renewable sources like wind and solar, and also in the widespread use of power electronics to connect these sources to the power system. Time varying renewable penetrations have been steadily increasing and some countries like Denmark and Ireland regularly operate with more than 50% of generation coming from such sources. Many reports have been published covering the need for changes in the structure of power systems to accommodate such increasing penetrations. A number of reports have been published in recent years assessing the potential for power systems to absorb high penetrations of renewable power generation. Some, like the 2011 report produced for the UK's Committee on Climate Change [22], covering the technical constraints on renewable generation up to 2050, and the NREL 2012 Renewable Electricity Futures Study [23], are country specific. Others are wider in scope, such as Integration of Renewable Energy in Europe [24] and the International Energy Agency (IEA) report: The Power of Transformation (wind sun and the economics of flexible power systems) [25], both published in 2014. These reports draw on the results of modelling studies (based on hourly energy balancing) for a range of potential future scenarios. They are pretty consistent in their conclusions and recommendations; these are briefly summarised here.

- Increasing renewable energy penetration need not result in significant increases in overall system generation costs
- Very high renewable energy penetration (up to 90%) is possible
- Transmission capacity must be substantially increased
- Flexible conventional (back-up) generation plant must be installed
- Mixing different renewable energy sources is beneficial
- The distribution system may need substantial upgrading
- Demand-side management (utilising the expected increase in heat pump and electric vehicle loads) is essential
- Dedicated energy storage remains too expensive for wide-scale use; however, this may change as the electric vehicle revolution promotes the development of novel and cheaper batteries
- Renewable energy sources should contribute to ancillary system services (such as frequency response)
- Renewable power generation forecasting is important and should be improved
- Market design must be improved to deliver coherent system operation

None of these conclusions should come as a surprise to readers of this book, indeed many follow naturally from the material presented in earlier chapters. Because they are based on hourly modelling, the dynamic challenges, resulting, for example, from low inertia, are not addressed. Moreover, the reports don't deal in any detail with the procedures that power system operators will need to adopt to deliver reliable and effective supply. The remainder of this chapter will outline the key developments required to facilitate a sustainable electricity supply system. These comprise: transmission development; active distribution systems (including demand-side management); operation of low inertia power systems and ubiquitous power electronics.

8.7.1 Transmission System Evolution

The European wind DC Super grid concept introduced in Section 8.2.2 is one proposal for how power systems might become increasingly interconnected in the future. Such improved and extensive interconnection will facilitate geographical aggregation and time smoothing of renewable energy generation. Analysts also argue that it extends

Figure 8.17 European and north African super grid for widespread renewable energy interconnection. Source: https://commons.wikimedia.org/w/index.php?curid=5443148.

the potential for market trading. Further extending the domain to include north Africa and including other renewable sources is central to the Medgrid and DESERTEC concepts. Combined, these are represented in Figure 8.17. Some have even speculated that the entire planet should be connected up so as to allow solar power to be transferred from areas experiencing daytime, and thus generating, to areas in darkness but requiring electricity [26]. Reference [27] proposes a methodology for selecting sites for solar power generation that when interconnected, and connected to load centres, maximises the proportion of the load that can be met in this way. The authors have shown that by linking regions in different time zones and on the two hemispheres solar intermittency can be substantially dealt with. A number of examples are given including: a Pan-American assembly of six sites (Chihuahuan desert, Mojave Desert, Sonoran Desert, Atacama Desert, a site in Argentina adjacent to the Atacama and Bolivia Litoral); a group of four sites from China and Mongolia (East and Easternmost Gobi desert), Ordos (autonomous area of Inner Mongolia) and Sukhbaatar (Mongolia) and eight sites across southern Africa (2 sites in Botswana, two sites in Namibia, three sites in South Africa and one in Tanzania). More detailed analysis of the Pan-American system proposal can be found in Reference [28], an alternative Asian system combining China, Mongolia and Australia is discussed in Reference [29]. Although these studies do include estimates of the required infrastructure investment, in practice it is highly unlikely that the costs of such transmission corridors would be affordable and there are also significant political obstacles.

Whether or not the extensive network connections outlined previously turn out to be feasible and economically attractive, it is clear that in Europe and elsewhere that transmission expansion is desirable. Europe has already seen a steady growth in HVDC connections, and considerably more is planned, as is clear from Figure 8.18. And interconnection is not confined to HVDC; conventional AC transmission reinforcement is

Figure 8.18 Existing and planned HVDC connections in Europe. Source: https://commons.wikimedia .org/w/index.php?curid=5553728.

commonplace. Not all commentators on future power systems agree with increased interconnection; some argue that with increased distributed (or dispersed) generation and smart grid technology, loads will primarily be met by local generation with a reduced need for bulk transmission. These ideas will be explored later in this chapter.

8.7.2 Low Inertia Power Systems

Time variable renewable energy sources, in particular wind and PV are connected to the power system using power electronics, and thus contribute no physical inertia. With the steady increase in penetration of such sources, power system operators have become concerned that this can lead to system instabilities.

The ENTSOE (European Network of Transmission System Operators for Electricity) has outlined a response to the challenge of low inertia power systems; this includes:

- Methods and tools to ensure the required level of inertia to the transmission system, and connection of relevant equipment to the networks.

- Identification of possible links between the electricity system and the other energy systems in the specific view of help.
- Increasing inertia (real or synthetic).
- Deployment of Wide Area Control (WAC) devices at the pan-European, system-wide level, which will enable the operators to operate the system close to its stability margins without jeopardising its security.

A number of studies of low inertia power systems have been published in recent years. For example, in Reference [30], based on simulation modelling of a power system with a high wind penetration, it is concluded that:

- Poor frequency stability resulting from the reduction of system inertia is a serious concern in the development of the low-carbon power system;
- The online synchronous generator fleet needs a better coordination to accommodate the variable wind power;
- Frequency stability should be seriously studied for the development of the low-carbon electric power system with the high penetration of renewable energy sources, especially the initial frequency response in the system following a power imbalance.

A detailed study of the impact of high penetrations of inverter-based generation that focused on the dynamic control of these sources, see Reference [31], concluded that new modes of power system oscillation were created, some involving the synchronous generators and some purely between different inverters. The characteristics of the latter were dependent on the nature of the converter controllers and this suggests challenges for such system modelling as the controller specifications will not always be known, and there will be such large numbers of such generators in typical future power systems.

A recent PhD project from University College Dublin [32] examined the issue of plant scheduling in low inertia power systems. Different approaches to ensure that sufficient inertia is connected to the power system at any time so as to ensure system stability have been assessed in terms of their economics. Simple must-run constraints on synchronous plant were found to be ineffective.

It is also clear from numerous simulation studies that synthetic inertia from wind and PV, as introduced in Chapter 3, can make substantial contributions to frequency control. If these are coupled with frequency contributions from dynamic load management (from water heating, space heating and electric vehicle charging loads), it should be possible to operate power systems with potentially no physical inertia. Rates of change of frequency (ROCOF) would probably increase. The most difficult circumstances will probably be when there are very few synchronous generators connected to the system, rather than when there are none. This is because the higher ROCOF will pose challenges to the stable connection of those few synchronous generators. Some countries have plans to build synchronous condensers (unloaded synchronous machines), or to create these from decommissioned power stations, to directly add inertia to the system.

8.7.3 Distribution Network Evolution

Traditionally, engineers have regarded electricity distribution networks primarily as passive conduits of power from the high voltage transmission system to the loads and equipment has been installed on a 'fit-and-forget' basis. However, some active control elements are now being introduced as far down as the 11 kV network for fault restoration to improve customer supply reliability. Some limited voltage control is also incorporated as described in Chapter 6, but these networks can generally be viewed as passive in operation. All this is changing with the growth of distributed generation.

The connection of smaller combined cycle gas turbines (CCGTs) at distribution voltages has been increasingly common in the European community over the last 15 years and similar trends are occurring elsewhere. A significant proportion of the UK's wind energy capacity, especially the older smaller wind farms, were connected at 11 kV and so embedded in the distribution system. In the USA, distributed generation consists primarily of standby diesel generator sets that add to system security by operating at times of high network stress. In addition, there is coming on to the market a wide range of very small generators (micro-CHP, PV and micro wind) that are designed for use in domestic and small commercial premises. These trends are expected to continue. As regional penetrations rise new ways of designing and operating distribution networks will evolve. Distribution Network Operators (DNOs) and researchers around the world are presently investigating these future scenarios and coming up with technological solutions to the operational challenges that can in the end deliver reliable power supplies at reduced cost.

Changes to be expected will include active and automatic network reconfiguration schemes aimed at facilitating high penetration of distributed generation while reducing customer minutes lost due to faults. Monitoring of the system will be much more extensive and will include data from smart metering installed throughout the system, down to individual household level. This will open up the possibilities of demand-side management of loads so as to better accommodate time varying renewable power generation. As part of these changes, it is anticipated that DNOs will become Distribution System Operators (DSOs) as they take on wider responsibilities for managing the system and delivering services to the system operator. A key factor that will distinguish such future distribution systems from those of the past is the massively increased use of digital communications technology; such systems are now referred to as 'smart grids'. Specific aspects of such networks are discussed in more detail next.

8.7.3.1 Active Networks

The network above the 11 kV system is actively controlled with activity increasing at higher voltage network layers. For example, the 30/10 kV tap-change transformer in Figure 5.11 is fitted with an automatic voltage controller that monitors the low voltage level and adjusts the taps to keep it within specified limits. Higher up in the chain, at the transmission level, not only active but also reactive power flows are appropriately regulated down transmission lines to ensure that the voltage profile is acceptable, the system is secure under contingencies and that it operates as economically as possible. Additionally, under fault conditions, reclose circuit breakers are controlled to improve reliability of supply.

Active control involves some or all of the following functions performed on a continuous basis:

- the measurement of system parameters, for example V, I, P, Q, f and network topology;

- comparing these measurements with set values or with desirable values obtained from analytical computations of the network model, for example load flow analysis, contingency analysis, optimum economic dispatch and so on;
- activating a control element to implement the desirable changes; for example, increasing the excitation of or the steam input to a synchronous generator, or opening or closing a transmission line;
- checking that the changes have been implemented and warning control engineers if the control actions fail.

Distribution networks are extremely extensive. For example, the UK has over 290 000 km of overhead and 470 000 km of underground cable. The conversion of this mostly 'passive' network into an 'active' network is a major task and will take place gradually as more distributed generation is adopted. A salient fact is that distribution networks in developed counties have been in place for several decades and will require updating soon in any case due to ageing. This creates a unique opportunity to enforce a culture change and design into the new networks the capability of embracing embedded generation and demand-side management rather than simply repeating yesterday's technology.

In the UK, Ofgem, the Office of Gas and Electricity Markets, presently supports distribution system innovation through the Network Innovation Allowance. These allow DNOs to explore and demonstrate new approaches and technologies and will assist them in making the transition to DSOs. Projects can cover low-carbon technologies like electric vehicles and heat pumps, low-carbon generation like PV, network operation including IT, as well as energy storage and demand response. Many counties offer such support, often under the heading of smart grids.

8.7.4 Problems Associated with Distributed Generation

The introduction of embedded generation into distribution networks will have a number of technical implications in the following areas.

8.7.4.1 Fault Levels

In urban areas the distribution networks are designed to have as high a short-circuit level as possible. This is helpful in minimising the effect that one consumer has on another and in maintaining the consumer voltage as close as possible to the nominal level. Economics dictate that distribution transformers, circuit breakers and cables are rated as near to the maximum load they are likely to transmit as possible. This small margin between operation and rating implies that if embedded generation is installed, the short-circuit level may increase beyond the plant capabilities.

8.7.4.2 Voltage Levels

One feature of distribution radial circuits is that they supply a number of distributed consumers; economics dictate that they taper in size and therefore current carrying capacity along their length. Embedded generation connected to the end of a long rural circuit is likely to increase the local voltage above statutory levels. Increased demands due to electric vehicle charging and the installation of domestic heat pumps will have the opposite effect so there is an opportunity to limit voltage excursions through the application of demand-side management.

8.7.4.3   Network Security

Connection of embedded generation has to comply with planning standards that aim to maintain the supply security at the pre-connection level. This may adversely limit the size and type of the embedded generator. The possibility exists that under fault conditions where the supply from the grid is lost, the local system is run in island mode fed solely from the embedded generator(s). In this case, embedded generation will enhance security. This mode of operation requires local active control and is addressed later.

8.7.4.4   Network Stability

Under certain fault conditions, the system dynamics are excited and it is possible that the characteristics of an embedded generator are such that the ensuing oscillations may cause tripping of the local network. If the dynamics of the generator are known, a stability study can be carried out before connection and if instabilities are detected, control systems theory can be applied to devise stabilising networks.

8.7.5   Options to Ameliorate the Technical Difficulties

8.7.5.1   Planning Standards

Present planning standards are based in the main on deterministic principals, in effect designing for the worst-case scenario, and therefore are highly conservative. New standards could be developed that rely on the probabilistic nature of renewable energy and of demand patterns. Such standards would be more forgiving and reduce the connection costs and penalties imposed on embedded generation. Dynamic line rating, as described in Chapter 5, Section 5.10, is a good example that takes into account ambient conditions and estimates current carrying capability of a line in a probabilistic manner.

8.7.5.2   Using Power Electronics Technology

High power electronics have made enormous advances over the last decade. These advances have resulted in new devices capable of increasing the flexibility of AC power systems. Such devices, known collectively as Flexible AC Transmission Systems (FACTS), offer considerable opportunities and challenges to power systems engineers. The challenges stem from the necessity of extending as far as possible the capacity of existing transmission lines without affecting adversely the voltage profile and jeopardising stability. Building additional transmission facilities would have solved some of the problems but these are often limited by economic and environmental considerations. FACTS covers a range of electronic based systems such as Static VAr compensators, thyristor controlled capacitors, phase-shift transformers, inter-phase power controllers and unified power flow controllers. Some of these devices have been used to assist in the integration of RE sources and are likely to facilitate future connections.

As has already been described in some detail, the nature of RE sources is such that they often require a power-electronic interface, usually a pulse-width modulation (PWM) converter, for connection to the grid. This interface could be regulated, not only to utilise optimally the renewable energy resource, but also to make the generator 'connection friendly'. The local or perhaps some other critical nearby voltage can be monitored and regulated by reactive power injection/extraction. The active power injected into the node may be limited at times to safeguard other local operating restrictions or requirements. The interface could be made sensitive to local faults so that it limits the contribution made by the local generator but does not limit it to the extent of endangering network stability. Finally, it could sense the loss of mains and arrange that the local network operates as an island arranging synchronisation and reconnection when the mains is re-established. Such converters are considered as grid supporting, since they rely on other generators in the system to control system frequency, and to an extent voltage. Converters can also be controlled to be grid forming, setting the grid voltage and frequency in the absence of other generation sources. This would allow the operation of AC microgrids with no synchronous generators. An excellent review of the different modes of converter operation and their control can be found in Reference [33]. Microgrids are discussed in more detail later.

Such extended local control activity will require substantial information processing and transfer, key characteristics of the smart grid. Interactive local control systems could be developed so that the embedded generator interface is in communication with other similar local units such as FACTS and with the local tap-change transformers, so that the most appropriate control actions are taken. The required communications may be provided through the Internet.

8.7.5.3 Islanding

An actively managed distribution network with locally generated power that approximately balances the demand could theoretically be run autonomously if connection to the mains were to be lost due to a fault, operating in effect as a microgrid (see later). Such operation would require suitable controlled generation and sophisticated local control actions to first maintain the frequency close to the nominal value and then to reconnect the islanded system back to the mains when the fault is cleared. The capability to island could increase the reliability of supply to the local consumers but at the cost of more complex software and hardware. The desirability of such operation depends crucially on the typical reliability of supply experienced by the consumers. If the reliability is high, say one short interruption per year on average, the benefits of islanding may be doubtful. There are also concerns that collapse of the wider network may occur due to the loss of contributions from the islanded system.

8.7.5.4 Dynamic Loads

In Chapter 3, Section 3.8.3, an application of dynamic load control was discussed as a means of regulating frequency; that is, providing power balance in a power system. In such an application, the first requirement is to be able to measure the power system's frequency accurately. A controller needs a cost-effective way to measure the system frequency by, say, using a simple microcontroller to count the number of processor cycles in a full AC cycle. The controller then needs to be aware of the needs of the appliance. In the case of a refrigerator, this means simply measuring the internal air temperature. A relay or other switching device is then added so the controller can switch the compressor (in the case of a refrigerator) or the heating element in the case of an immersion heater. The approach can be extended to a variety of loads: domestic water heating; electric heating and cooling and electric vehicle battery charging. There are a number of published studies demonstrating the potential to control these loads without significantly disrupting the service required by the customer.

The application of such dynamic load control addresses the totality of system power balance. It is of course possible also to use this technology to service local distribution network requirements; for example, voltage as mentioned before. The totality of the dynamic loads at the end of a feeder supplying a large group of houses may be diverted from system frequency control to local needs; for example, matching from instant to instant the output variations of local embedded generation or assisting the maintenance of system security. Although much research on the potential of dynamic load control to contribute to system frequency stability has been completed, there is little evidence of any of this being put into practice. A particular challenge for the dynamic control of key domestic loads is the small economic value associated with individual households. Furthermore, system operators are also concerned that the response is not deterministic as with contracted response from large-scale generators. In reality, this may not be a problem in that the numbers of controlled loads would be so large that aggregate response should be able to be accurately predicted. Imaginative offerings to householders from aggregators, or possibly DSOs or energy suppliers, that provide suitable rewards for participation are required if the benefits of dynamic control are to be realised.

8.7.5.5 Demand-Side Management of Loads

In contrast to dynamic load control, demand-side management (sometimes called demand response) refers to the retiming of loads so as avoid periods of peak load when electricity costs are high, to absorb as far as possible surplus renewable energy generation and to mitigate high or low distribution system voltages. As far as possible, demand-side management tries to match loads to generation availability. It introduces essential flexibility into a power system and in particular supports the integration of high penetrations of time variable renewable power generation. The concept of switching loads to mitigate peaks in demand is not new. For example, off-peak tariffs for switched night-time loads have been used UK for many decades, commonly used with storage heaters, and in the USA since the 1970s domestic air-conditioning loads have been switched off through central control to decrease peak loads on the system. What has changed is that, with the roll out of smart meters, power system operators can get a far more detailed understanding of the load, also any generation within the home and its spatial location, and consumers can receive information on time varying tariffs. This opens up the possibility for consumers to respond to changing levels of renewable generation, as well as the overall system load, and by doing so reduce their electricity bills, while at the same time contributing to easing power system operation. Recent reviews of demand response can be found in References [34, 35].

Demand-side management is a topic that has been discussed for a long time in power system circles but its considerable potential has not yet been exploited to any significant extent. In part, this is because electricity suppliers have yet to offer time variable or time-of-use (ToU) tariffs that would incentivise consumers to participate in such schemes, and market structures have not been developed to give financial reward for delivering services. An interesting Reference, [36], which reviews the 'socio-techno-economic' barriers to implementation, highlights a wide range of technical, market and behavioural issues that need to be overcome. The researchers conclude that a central factor is the complexity of demand response and the interaction of the many issues involved.

8.7.5.6 Storage

There is at present considerable interest in the possible applications of energy storage in power systems. The technologies are diverse and at different stages of development as discussed in some detail in Chapter 3. They range from a variety of batteries, high speed flywheels, supercapacitors and regenerative fuel cells. Recent improvements in battery technology, driven by electric vehicle applications in the main, have encouraged many to believe that they will soon be extensively deployed. Nevertheless, batteries remain expensive and there will be environmental issues associated with their widespread use. Despite this many governments are enthusiastic about their potential and are exploring changes in regulation that should make it easier for battery storage systems to participate in electricity markets. A related possibility is so-called 'second use' where partly degraded batteries that have been used in electric vehicles but can longer sustain the required range, are re-deployed to form large grid-connected energy storage units.

Local energy storage would assist embedded generation from renewable energy by providing a buffer between the variability of supply and demand. As the energy storage device would be more than likely to be interfaced to the mains through a power-electronic converter, the capabilities of such a converter would also be available to assist the local network.

At the limit, the extensive uptake of embedded generation in the domestic area, such as PV, fuel cells and micro-turbines combined with improved efficiency of domestic appliances and levels of insulation, could make homes self-sufficient in terms of energy. If such technologies were combined with energy storage, the nature of the distribution network could decline to a minimum infrastructure system providing only occasional energy delivery. The idea of energy self-sufficient homes where the peak to average electricity demand ratio is as high as 10–20 (Figure 3.1) is unlikely to be cost effective compared to the substantial improvement that can be achieved in that ratio by grouping several houses together (Figure 3.2). Such small groupings of consumers that benefit from the averaging effect of aggregation is a more likely candidate for future development; this approach is now widely referred to as a 'microgrid'.

8.7.5.7 Microgrids

Microgrids may consist of a small community, a housing estate, a university, a commercial or industrial area or a municipality, or a section of a larger power network that can be islanded when the need arises. The key concept is that it includes a number of consumers and a number of small generators (often referred to as micro-generators), located in close proximity so that line losses are minimised. The totality of the installed capacity in a microgrid must be capable of supplying the maximum local demand. Microgrids promise substantial environmental benefits by facilitating the integration of renewables such as PV arrays and small wind turbines. The microgrid concept is heavily reliant on small scale generators, often micro-CHP generators that are mostly driven by the production of heat during winter, and PV generators that provide electrical energy during summer. Some storage capability is also assumed. The correct mix is obviously important but would be facilitated by the diversity of the consumer group. The CHP is most commonly provided by diesel engines, so such systems can never be truly sustainable. Nevertheless, the topic has attracted considerable research and demonstration over the last 15 years, and International Standards are underdevelopment such as IEC/TS 62898-1, Guidelines for General Planning and Design of Microgrids, and

IEC/TS 62898-2, Technical Requirements for Operation and Control of Microgrids. A useful review that sees microgrids as a means to facilitate distributed generation and active distribution networks can be found in Reference [37]. Microgrids that form part of a larger power network, sometimes called utility microgrids, tend to be considerably larger in size, both in terms of rating and geographical extent. They can also involve a much large number of generators that can be more varied in type than for stand-alone microgrids that might have only one generation source. This poses challenges in the modelling and control to deliver stable operation. Reference [38] assesses the range of possible stability issues for microgrids and their causes, and also provides an appraisal of different approaches to improving stability, including stabilisers that can be added to VSCs, adding flywheel energy storage and the use of DSTACOMs (distributed static compensators).

8.7.5.8 Virtual Power Stations

Virtual power stations are a relatively new concept. A virtual power plant is defined by Wikipedia as a cluster of distributed generation installations (such as micro-CHP, wind turbines, small hydro, back-up gensets etc.) that are collectively run by a central control entity. Concerted operation of this plant can in theory provide network services such as power balancing and frequency control. The concept is a response to market mechanisms that penalise the variability and lack of predictability of small individual generators, particularly those based on renewable energy sources. However, since these aggregations are expected to reflect commercial arrangements that might involve highly dispersed plant, rather than geographically specific clusters, the operation of such virtual plants might conflict with the requirements for local power and voltage control. These issues are the topic of ongoing research and, so far, there is no clear consensus on the value of this approach.

8.8 Conclusions

National and local electricity grids, metering systems and regulatory arrangements that were created for a world of large-scale, centralised power stations are now in the process of being restructured to support the integration of increasing amounts of renewable power generation.

Intermittency is not a 'flaw' or shortcoming as traditional 'reliability' concepts imply. On the contrary, requiring a system to always deliver generation that matches a fleeting peak load gives rise to a set of generation and network assets that are invariably drastically overspecified and underemployed, a situation long overdue for change through innovative policy development.

When a new generation technology, wind for example, is 'shoe-horned' into an existing system that evolved to support technology from a previous era, things do not automatically work correctly. To the system operator it may seem clear that the new technology needs to change so that it takes on the characteristics of conventional generators (i.e. thermal plant with synchronous generators and physical inertia). Although this can be achieved to an extent as explained at a number of points in this book, this may not always be the most economic approach. For example, the derating of wind or PV plant so as to create the headroom required to deliver droop response may simply waste potentially useful generation. A much more comprehensive system approach is required. Increased flexibility is important, but this can be delivered by a range of approaches, not least by making the electricity load, traditionally treated as a given, into something flexible. If technological potential is to be effectively exploited, it is the underlying system and the way that it is operated that needs to metamorphose.

Compared to the breath-taking speed of developments in IT, the technology of power generation and distribution has developed too slowly; the major development having been with renewable power generation, in particular wind on and offshore, and PV. One reason is that the natural lifecycle of asset renewal in power systems is measured in decades rather than years. In truth, the vast range of IT technology has to date barely affected the electricity networks. Although smart meters have been rolled out in many countries, their potential is not being exploited. They are mainly at present simply used for remote meter reading. Interestingly, the long-delayed revolution required is being precipitated by the growing importance and impact of renewable sources. The distributed and variable nature of renewables requires intelligently controlled power-electronic hardware to maintain power quality and optimise energy flows and this technology is now mature. This required revolution in flexibility and controllability of the network will, of course, be of immense benefit to traditional generation as well. Much of the distribution network in most countries is at least 40 years old and will have to be replaced over the next decade. There is therefore a great opportunity to introduce innovative technologies rather than replace like with like.

Traditionally, power systems have been run on the policy that whenever the consumer demands energy this should be instantly satisfied from dispatchable generation. With increasing proportions of renewable generation this philosophy must be changed but without substantial inconvenience to the consumer. Demand-side flexibility will require that consumers adjust their demand profile to meet supply through deferrable loads. This will require intelligent control and could be done through time-of-use tariffs requiring customer voluntary response, or alternatively hard-wired direct control or variations of these approaches. If demand can shift time-wise to suit generation, the availability of primary energy will be dictating the pattern of consumption rather than the opposite. In such a power system, 'base load' electricity may not continue to have the significance it has at present. As already indicated, implementing such approaches is far from straight-forward and much effort needs to be dedicated to overcoming the barriers. The expected growth of EVs will create enormous opportunity in this regard.

An active distribution network is by definition more complex than the current system, and building and managing it will require new skills. In this future power system, renewable generation will be injected and active control will be implemented at all system levels.

To conclude, a fundamental rethink will be needed of the way electricity is generated, transmitted and distributed in the power systems of the future. Variable generation infeeds coupled with bidirectional network flows will become commonplace. This rethink not only requires the adoption of new technologies but also a change in the mindset or 'culture' of power system engineers. Extensive research has been completed highlighting new ways of working and the subject remains a focus point for academic research. The challenge is to implement this new knowledge extensively through power systems. Commercial power system engineers tend to be conservative and limit the rate of change to their systems; this caution reflects legitimate concerns to maintain system reliability. However, the key driver of renewable power integration will reduce system reliability (or alternatively be wasted through curtailment) unless new modes of system operation are adopted. This book has endeavoured to cover the basic principles of how power systems operate, and also importantly how they can evolve from the present time when renewable energy contributions are limited albiet growing rapidly, towards a future when they become the dominant form of generation.

References

1 Jamieson, P. (2011). *Innovation in Wind Turbine Design*. Wiley.
2 Zhang, J., Fowai, I., and Sun, K. (2016). A glance at offshore wind turbine foundation structures. *Shipbuilding* 67 (2): 101–113.
3 ABB Group, ABB Special Report: 60 Years of HVDC, ABB report, available at https://library.e.abb.com/public/aff841e25d8986b5c1257d380045703f/140818%20ABB %20SR%2060%20years%20of%20HVDC_72dpi.pdf (accessed May 2019), 2014.
4 Jacobson B., Jiang-Hafner Y., Rey P. et al. 'HVDC with voltage source converters and extruded cables for up to ±300 kV and 1000 MW', CIGRE Paper B4-105, 2006.
5 Wensky D., FACTS and HVDC for grid connection of large offshore wind farms, EWEC 2006, Athens, 27 February-2 March 2006.
6 Lian Y. and Finney S., DC Collection Networks for Offshore Generation; Proc. IET RPG Conference, 2013.
7 Elliot D.W., Finney S.J. and Booth C., Single converter interface for a cluster of offshore wind turbines. In, Proceedings of IET Conference on Renewable Power Generation, Edinburgh, 2011.
8 Adam, G.P., Gowaid, I.A., Finney, S.J. et al. (2016). Review of DC-DC converters for multi-terminal HVDC transmission networks. *IET Power Electron.* 9 (2): 281–296.
9 O'Connor, E. (2006). European unity, a vision for sustainable power in Europe. *Renewable Energy World*: 124–127.
10 Tan, A., Lin, X., Sun, J. et al. (2016). A novel DFIG damping control for power system with high wind power penetration. *Energies* 9 (521): 1–15.
11 Edrah, M., Lo, K.L., and Anaya-Lara, O. (2016). Reactive power control of DFIG wind turbines for power oscillation damping under a wide range of operating conditions. *IET Gener. Transm. Distrib.* 10 (15): 3777–3785.
12 Singh, M., Allen, A.J., Muljadi, E. et al. (2015). Interarea oscillation damping controls for wind power plants. *IEEE Trans Sustainable Energy* 6 (3): 967–975.
13 Jiang, J.N., Tang, C.Y., and Ramakumar, R.G. (2016). *Control and Operation of Grid-Connected Wind Farms*. Springer.
14 Kanev S., Savenijie F., Soleimanzadeh M. and Wiggelinkhuizen E. Wind farm modelling and control: an inventory. ECN report ECN-E-13-058, December 2013.
15 Campagnolo, F., Petrović, V., Schreiber, J. et al. (2016). Wind tunnel testing of a closed-loop wake deflection controller for wind farm power maximization. *J. Phys. Conf. Ser.* 753: 032006.
16 Infield, D. and Staffell, I. (2015). Micro-wind turbines for domestic applications. In: *Chapter 8 in Domestic Microgeneration: Renewable and Distributed Energy Technologies, Policies and Economics* (eds. I. Staffell, D.J.L. Brett, N.P. Brandon and A.D. Hawkes), 183–199. Earthscan.

17 Bazilian, M., Onyeja, I., Liebreich, M. et al. (2013). Re-considering the economics of photovoltaic power. *Renewable Energy* 53: 329–388.

18 Gagnon, L., Berlanger, C., and Uchiyama, Y. (2002). Life-cycle assessment of electricity generation options. *Energy Policy* 30: 14.

19 Polman, A., Knight, M., Garnett, E.C. et al. (2016). Photovoltaic materials: present efficiencies and future challenges. *Science* 352 (6283).

20 Demirbas, A. (2005). Potential applications of renewable energy sources, biomass combustion problems in boiler power systems and combustion related environmental issues. *Prog. Energy Combust. Sci.* 31: 171–192.

21 Molino, A., Chianese, S., and Musmarra, D. (2016). Biomass gasification technology: the state of the art overview. *J. Energy Chem.* 25: 10–25.

22 Analysing Technical Constraints on Renewable Generation to 2015, A report to the Committee on Climate Change, March 2011, Poyry.

23 NREL, Renewable Electricity Futures Study; NREL report, December 2012.

24 Integration of Renewable Energy in Europe (final report on EU project ENER/C1/427-2010, June, 2014.

25 International Energy Agency (2014). *The Power of Transformation – Wind, Sun and the Economics of Flexible Power Systems.* IEA.

26 Chatzivasileiadis, S., Ernst, D., and Andersson, G. (2013). The global grid. *Renewable Energy* 57: 372–383.

27 Grossmann, W.D., Grossmann, I., and Steininger, K.W. (2013). Distributed solar electricity generation across large geographic areas, part I: a method to optimize site selection, generation and storage. *Renewable Sustainable Energy Rev.* 25: 831–843.

28 Grossmann, W.D., Grossmann, I., and Steininger, K.W. (2014). Distributed solar electricity generation across large geographic areas, part II: a pan-American energy system based on solar. *Renewable Sustainable Energy Rev.* 32: 983–993.

29 Taggart, S., James, G., Dong, Z.Y., and Russell, C. (2012). The future of renewables linked by a transnational Asian grid. *Proc. IEEE* 100 (2): 348–359.

30 Wu, D., Javdi, M., and Jiang, J.N. (2015). A preliminary study of impact of reduced system inertia in a low-carbon power system. *J. Modern Power Syst. Clean Energy* 2 (1): 82–92.

31 Quintero, J., Vittal, V., Heydt, G.T., and Zhang, H. (2014). The impact of increased penetration of converter control-based generators on power system modes of oscillation. *IEEE Trans. Power Syst.* 29 (5): 2248–2256.

32 Daly P. "Scheduling low inertia power systems." PhD thesis. University College Dublin, 2017.

33 Rocabert, J., Luna, A., Blaabjerg, F., and Rodriguez, P. (2012). Control of power converters in AC microgrids. *IEEE Trans. Power Electron.* 27 (11): 4734–4749.

34 Bosmann, T. and Eser, E.J. (2016). Model-based assessment of demand-response measures – a comprehensive literature review. *Renewable Sustainable Energy Rev.* 57: 1367–1656.

35 Esther, B.P. and Kumar, K.S. (2016). A survey on residential demand side management architecture, approaches, optimization models and methods. *Renewable Sustainable Energy Rev.* 59: 342–351.

36 Good, N., Ellis, K.A., and Mancarella, P. (2017). Review and classification of barriers and enablers of demand response in the smart grid. *Renewable Sustainable Energy Rev.* 72: 57–72.

37 Lopes, J.A.P., Madureira, A.G., and Moreira, C.C.L.M. (2013). A view of microgrids. *Wiley Interdiscip. Rev. Energy Environ.* 2 (1): 86–103.

38 Majumder, R. (2013). Some aspects of stability in microgrids. *IEEE Trans. Power Syst.* 28 (3): 3243–3252.

Appendix A

Basic Electric Power Engineering Concepts

A.1 Introduction

This appendix covers some of the very basic topics that underlie the generation, transmission and use of electrical energy, but it is not intended to be a substitute for a good textbook on basic electrical engineering. For readers who have no background whatsoever in this topic, selective study of relevant chapters in a typical good textbook as, for example, Reference [1], would be essential.

This appendix is intended for readers who have studied 'electricity' as a subsidiary subject at university and need to refresh their memories. It is likely that the 'electricity' they have met was slanted towards the understanding of electronic circuits and devices. This is because the major concern of electrical engineering since the invention of the transistor and of the computer has been the 'communications' and 'control' rather than the 'energy' aspects of the discipline. It is, therefore, possible that, although such readers are knowledgeable in some aspects of electricity, they are unfamiliar with the ideas and conventions developed over the years by power system engineers. If this is the case, they are likely to find this appendix useful.

For a power system to be an efficient channel of energy flow from generators to consumers, the task should be accomplished with the minimum energy loss. Because of the need to supply a constant voltage, it is also necessary that the desirable energy transfers are achieved within specified upper and lower voltage limits at the system junctions or nodes. These requirements have led to certain analytical concepts that are particular to power systems engineering.

A.2 Generators and Consumers of Energy

A start is made with the definition of the conventions used when power is to be calculated in electric circuits. In Figure A.1, the box represents any single piece or collection of equipment found in a power systems network. In the analysis that follows, whenever lower-case letters are used for circuit variables such as voltage, current, and power, it will be taken that these refer to instantaneous values. The voltage v_{AB} describes the potential difference of terminal A with respect to terminal B at one instant. If the potential of A is higher than that of B then v_{AB} is positive. It follows that $v_{BA} = -v_{AB}$. Now i_{AB} will be defined as the current that flows from A to B *through* the element. If the current

Renewable Energy in Power Systems, Second Edition. David Infield and Leon Freris.
© 2020 John Wiley & Sons Ltd. Published 2020 by John Wiley & Sons Ltd.
Companion website: www.wiley.com/go/infield-renewable-energy

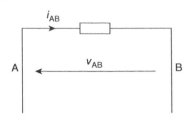

Figure A.1 Circuit element.

transfers positive charge from A to B then i_{AB} is said to be positive. Again, it follows that $i_{BA} = -i_{AB}$.

Power in electricity is given by the product of voltage and current. For the element in Figure A.1 the instantaneous power p is given by

$$p = v_{AB}i_{AB} \tag{A.1}$$

Figure A.1 encapsulates the convention for the reference voltage and current directions in this simple circuit configuration that, for obvious reasons, is known as a *one-port*. When setting up the reference directions in a one-port it is said that one looks into the terminals AB towards the circuit in question.

If v_{AB} and i_{AB} are both positive then p is also positive and the element is a *consumer* or *sink* of electrical power. The truth of this simple convention can be tested by examining the simple DC circuit of Figure A.2. Here, the circuit element is a plain resistor and terminals AB are connected to a battery of the polarity shown. With this DC *excitation* the instantaneous quantities are of course the same as the steady state values; v_{AB} and i_{AB} are both positive in this case and the power p associated with the resistor is consumed and irreversibly converted into heat. Figure A.3 is identical to Figure A.2 but now the battery can be considered as being the element in the box of Figure A.1. Therefore, the terminals AB are looked into downwards rather than upwards, with the result that current i_{AB} has reversed direction.

Figure A.2 DC circuit example.

Figure A.3 DC circuit example.

Current i_{AB} through the battery is now negative, v_{AB} is still positive and, therefore, Eq. (A.1) gives a negative value to the power associated with the battery. If a line is drawn at the interface AB, power is going into the interface if looked at towards the resistor, that is, the power is positive, but power is coming out of the interface if looked at towards the battery; that is, the power is negative. This, of course, satisfies the conservation of energy principle at the interface.

To sum up, the power associated with an element is taken as positive and therefore consumed when the current is flowing *through* the element from higher to lower potentials. If the reverse were the case, the power would be negative and the element would be a generator. All this is self-evident when dealing with DC circuits, but what would be the power associated with an element if v_{AB} and i_{AB} were continuously varying quantities such as sinusoids?

A.3 Why AC?

In some cases, AC holds no practical advantage over DC. In applications where electricity is used to dissipate energy in the form of heat, the polarity or direction of current is irrelevant. However, with AC it is possible to build electric generators, motors, and power distribution systems that are far more efficient or flexible than with DC, and so AC is used predominately across the world in high power applications. Section 4.2 provides information on how AC is generated through a rotating mechanical–electrical energy converter.

The relative simplicity of AC generators and motors translates into greater reliability and lower cost of manufacture. There is, however, an additional, very important advantage of AC over DC. If there are two mutually inductive coils and one coil is energised with AC, an AC voltage will be created in the adjacent coil. This device is known as a transformer and its mode of operation is described in some detail in Section 4.3. The fundamental significance of a transformer is its ability to step voltage up or down from the powered coil to the unpowered coil. As explained in Section 1.4, the transformer's ability gives AC an advantage unmatched by DC in the realm of power transmission and distribution.

A.4 AC Waveforms

In practice, AC generators are designed to produce a voltage that, to all intents and purposes, is a pure *sine wave*. There is a good reason why pure sine waves are used in power networks. A sine wave is the only waveform that, when differentiated or integrated, results in a replica of itself. As will be shown later, electric power system components, including transformers, respond differentially to a sine wave excitation and, thus, the shape of the waveform is maintained throughout the network. Additionally, three-phase electric motors are at their most efficient and produce a constant torque when fed from an AC supply of sine wave shape only. A sinusoidal voltage of frequency f can be written analytically as $v = \widehat{V} \sin \omega t$, where $\widehat{V}$ is the peak value of the sine wave and $\omega = 2\pi f$.

From now onwards, whenever AC is referred to, it is implied that the waveform is of sinusoidal shape. With AC, a measurement dilemma is encountered in expressing how large or small an AC quantity is. With DC, where quantities of voltage and current are generally stable, there is no trouble expressing what is the value of the voltage or current in any part of a circuit.

One way to express the magnitude (or *amplitude*) of an AC quantity is to measure its peak value $\widehat{V}$. However, when dealing with measurements of electric power, the best way to express the value of an AC waveform is to label it in terms of its ability to perform useful work. The way to do this is to assign to the waveform a 'DC equivalent' measurement, that is, to label it by the value in volts or amperes of the DC that would have produced the same amount of heat energy dissipation in the same resistance. The mathematical method of producing the DC equivalent value, known as root mean square (RMS), can be found in any elementary textbook on electricity. In this book, the RMS value of an AC voltage or current will be denoted by an italic capital letter, for example V, I. It can be shown that for a sine wave the RMS value is related to the peak by

$$V = \sqrt{2}\widehat{V} = 0.707\widehat{V} \tag{A.2}$$

Thus, the household voltage of 230 V has a peak of 325 V.

A.5 Response of Circuit Components to AC

Power networks consist mainly of generators, transmission lines, transformers, and consumers. Each one of these constituent parts consists, in turn, of a combination of three basic components, namely resistance R, inductance L and capacitance C. Additionally, generators have internally induced AC voltages caused by the interaction of moving conductors and magnetic fields. These electromagnetic actions are the mechanisms by which mechanical energy, conventional or renewable, is converted into electrical form. Uniquely, in photovoltaic systems the conversion of solar power is carried out directly into electricity without the aid of electromagnetics.

The AC source voltages are the drivers of all current flows in the network and, therefore, partly determine the distribution of power flows from the generators to consumers over the transmission network. The energy flows in a network consisting of traditional and/or renewable generators can be determined by the application of Kirchhoff's laws, but for this to be done the 'resistance' to the flow of AC current offered by the three basic components must be known. In other words, the relationship between AC voltage and current (Ohm's law) for each component must be determined.

A.5.1 Resistance

Ohm's Law for a resistance R in ohms is the linear relationship

$$i = \frac{v}{R} \tag{A.3}$$

where v in volts and i in amperes are the instantaneous values of voltage and current. Figure A.4 shows a plot of the applied voltage

$$v = \widehat{V} \sin \omega t \tag{A.4}$$

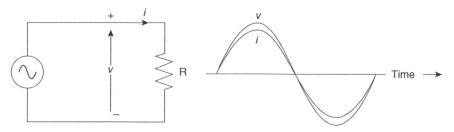

Figure A.4 Resistor excited by AC voltage.

and the resulting current i in a circuit consisting of an AC source and a resistor R. From Eq. (A.3) the resulting current is given by

$$i = \frac{\widehat{V}}{R} \sin \omega t \tag{A.5}$$

Equation (A.1) shows that the instantaneous power is given by

$$p = vi = \frac{\widehat{V}^2}{R}\sin^2 \omega t = 2\frac{\widehat{V}}{R}\sin^2 \omega t = \frac{V^2}{R}(1 - \cos 2\omega t) \tag{A.6}$$

where V is the RMS value of the AC voltage.

Equation (A.6) is plotted in Figure A.5. This is a double-frequency waveform of average value V^2/R and is fully displaced above the time axis. This confirms the evident fact that the power converted in the resistor from electrical into heat form is not dependent on the direction of current. It is concluded that to calculate the power in a purely resistive element any of the following expressions can be used:

$$P = \frac{V^2}{R} = RI^2 = VI \tag{A.7}$$

A.5.2 Inductance

Inductors do not behave like resistors. Whereas resistors simply oppose the flow of electrons through them (by dropping a voltage directly proportional to the current), inductors oppose *changes* in current through them, by dropping a voltage directly proportional to the *rate of change* of the current.

Ohm's law for an inductance L in henries is the following differential rather than linear relationship:

$$v = L\frac{di}{dt} \tag{A.8}$$

Figure A.5 Power in a resistive component.

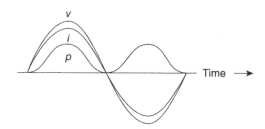

Assume that a sinusoidal current $i = \hat{I} \sin \omega t$ flows through an inductor of inductance L. Inserting this in the right-hand side. of Eq. (A.8) gives $v = (\omega L)\hat{I} \cos \omega t = \hat{V} \cos \omega t$, where

$$\hat{V} = (\omega L)\hat{I} \tag{A.9}$$

Figure A.6 shows the voltage/current relationship for an inductor. The voltage represented by the cosine curve is 90° ahead of the current sine curve. The voltage is now said to *lead* the current by 90° or the current to *lag* the voltage by the same amount. In general, if two AC waveforms are phase-shifted by 90°, it is said that these waveforms are in *quadrature* to each other. If the voltage is described by

$$v = \hat{V} \sin \omega t \tag{A.10}$$

then the current would take the form $i = \hat{I} \sin(\omega t - \pi/2)$, which through Eq. (A.9) becomes

$$i = \frac{\hat{V}}{\omega L} \sin\left(\omega t - \frac{\pi}{2}\right) \tag{A.11}$$

The instantaneous power is

$$p = vi = \hat{V}\hat{I} \cos \omega t \sin \omega t$$

and through Eq. (A.9),

$$p = (\omega L)\hat{I}^2 \sin \omega t \cos \omega t = (\omega L)I^2 2 \sin \omega t \cos \omega t = (\omega L)I^2 \sin 2\omega t \tag{A.12}$$

where I is the RMS value of the current.

Equation (A.12) is plotted in Figure A.7. Here the power is of double frequency, as in the case of the resistor, but it is wholly symmetrical with respect to the time axis. This means that a perfect inductor (one that possesses no winding resistance) is not associated with a net power transfer over the long run. During periods in the cycle when the

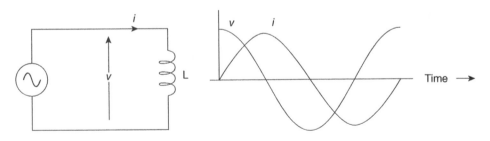

Figure A.6 Inductor excited by AC voltage.

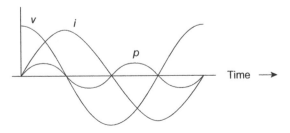

Figure A.7 Power in an inductive component.

power is positive, the inductor absorbs energy from the network to which it is connected and stores it in its magnetic field. During periods when the power is negative, the stored energy is returned to the rest of the system. It can be concluded that an inductor acts as a consumer or sink of electrical energy during part of the cycle and a generator of energy during the remaining cycle.

The purpose of power systems is to transfer electrical power from generators to consumers so that the latter could convert this energy usefully into heat, light, mechanical or chemical power or use it to drive IT systems. The power associated with an inductor serves none of these purposes. It consists of power oscillating between power system components.

A.5.3 Capacitance

Capacitors oppose changes in voltage by drawing or supplying current as they charge or discharge to the new voltage level (Figure A.8). The flow of electrons through a capacitor is directly proportional to the rate of change of voltage across the capacitor. This opposition to voltage change is a mirror image to the kind exhibited by inductors.

Ohm's law for a capacitance C in farads is the differential relationship

$$i = C\frac{dv}{dt} \tag{A.13}$$

If a sinusoidal voltage

$$v = \widehat{V} \sin \omega t \tag{A.14}$$

is applied to a capacitor, the resulting current from Eq. (A.13) is

$$i = (\omega C)\widehat{V} \cos \omega t = \widehat{I} \sin\left(\omega t + \frac{\pi}{2}\right) \tag{A.15}$$

where

$$\widehat{I} = (\omega C)\widehat{V} \tag{A.16}$$

The current is now said to *lead* the voltage by 90° or the voltage to *lag* the current by the same amount.

The simple capacitor circuit is characterised by a power curve (Figure A.9), which is similar to that of the pure inductor.

A capacitor, also a storage element, does not dissipate power as it reacts against changes in voltage; it merely absorbs and releases power, alternately.

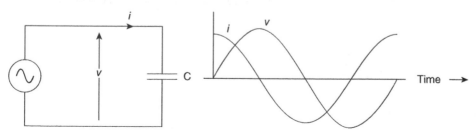

Figure A.8 Capacitor excited by AC voltage.

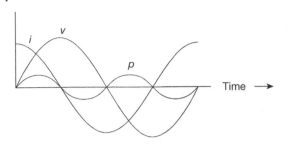

Figure A.9 Power in a capacitive component.

There is one final observation that must be stressed. A comparison between the sinusoidal power waveforms in Figures A.7 and A.9 reveals that they are always in opposition when viewed with respect to the voltage waveform. That is to say, during periods when the power in the inductance is positive, the power in the capacitance is negative and vice versa. This is an important fact that will be returned to later.

A.6 Phasors

In the previous analysis, the AC variations of voltage, current and power were depicted as sinusoidally varying functions of time. This is acceptable for an introductory exposition of the relationships of these quantities for single-circuit elements. In real life, power circuits consist of numerous combinations of such elements and a more appropriate tool is required for their analysis.

The kind of information that expresses a single dimension, such as linear distance or temperature, is called a *scalar* quantity. The voltage produced by a battery, for example, is a scalar quantity. However, when alternating current circuits are analysed, it is found that voltage and current are not the familiar one-dimensional quantities. Rather, these quantities, because they are dynamic (alternating in direction and amplitude), possess two dimensions, that is, amplitude and phase. Therefore, there is a need to work with mathematical techniques capable of representing two-dimensional quantities. A *complex number* is a single mathematical quantity able to express these two dimensions of amplitude and phase at the same time. A graphic representation of a complex number is known as a *phasor*.

It must be obvious that there must be some common frame of reference for angles to have any meaning. In this case, directly right in Figure A.10 is considered to be 0° and angles are counted in a positive direction going counterclockwise.

Figure A.10 The Argand diagram of complex numbers.

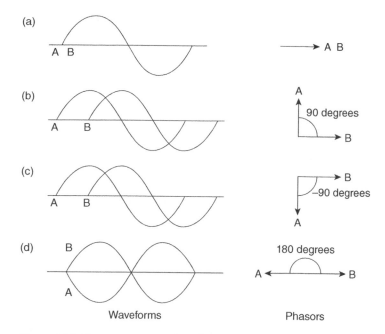

Figure A.11 Phasor representation of AC quantities.

The length of the phasor represents the *magnitude* (or amplitude) of the AC wave-form. The greater the magnitude of the waveform, the greater the length of its corresponding phasor. The angle of the phasor represents its *phase*, that is, the phase shift in degrees between the waveform in question and another waveform acting as a 'reference' in time. In this book, a phasor is denoted by a bold capital letter, for example *A*. Phasor *A* in Figure A.10 has a magnitude *A* and a phase angle *θ*. The representation of a phasor by magnitude and phase is known as its *polar* form. The polar form of phasor *A* in Figure A.10 is written as $A = A \angle \theta$. The phase of a waveform in a circuit is usually expressed with regard to the power supply voltage waveform (arbitrarily stated to be 'at' 0°). If there is more than one AC voltage source, then one of those sources is arbitrarily chosen to be the phase reference for all other measurements in the circuit. Phase is always a *relative* measurement between two waveforms rather than an absolute property.

In Figure A.11, A and B could represent the voltage and current waveforms, (a) for a resistor, (b) for an inductor, (c) for a capacitor and (d) could represent the antiphase relationship of power in an inductor and a capacitor.

A.7 Phasor Addition

If phasors that are not in-phase or antiphase are added, their magnitudes add up quite differently to that of scalar quantities. Figure A.12 shows an example of such an addition.

In fact, phasors add up in exactly the same way as vectors that represent velocities or forces in mechanical systems. As explained in Section A.4, the 'effective' value of an AC waveform is its RMS value; hence AC quantities, unless stated otherwise, are always

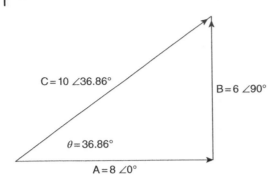

Figure A.12 Addition of phasors.

Figure A.13 Addition of AC voltages.

labelled in terms of their RMS value and are expressed by a capital letter. For example, V and I represent the RMS values of the general sinusoids

$$v = \sqrt{2}V \sin(\omega t + \phi) = \widehat{V} \sin(\omega t + \phi) \ \text{ and } \ i = \sqrt{2}I \sin(\omega t + \theta) = \widehat{I} \sin(\omega t + \theta)$$

respectively, where $\widehat{V}$ and $\widehat{I}$ are the peak values and ϕ and θ are phase angles with respect to a reference sinusoid.

If two AC voltages are added together by being connected in series as in Figure A.13, their voltage magnitudes do not directly add or subtract as with scalar voltages in DC. Instead, these voltages are complex quantities and, just like the previous phasors, which add up in a trigonometric fashion, an 8 V source at 0° added to a 6 V source at 90° (both RMS) results in 10 V at a phase angle of 36.86°.

The property of phasors that, although stationary on a page, enables them to represent AC time varying sinusoidal quantities is of inestimable value in the calculation of power flows in complex AC circuits containing voltage sources and combinations of R, L and C elements.

A.8 Rectangular Notation

Simple electric networks can be analysed through phasor diagrams. This is extremely tedious when a more challenging circuit is involved and becomes impossible for real life networks. Phasors can be represented mathematically by complex numbers and circuits can be analysed through algebraic manipulations without the laborious task of drawing phasors. There are two basic forms of complex number notation: the polar, described earlier, and the *rectangular* or *Cartesian*.

In the rectangular form, a complex number is denoted by its respective horizontal and vertical components. In order to distinguish the horizontal and vertical dimensions from each other, the vertical is prefixed with a lower-case j. The horizontal and vertical phasor components are known as real and imaginary numbers, respectively.

Figure A.14 shows the 'complex number plane' on which any phasor expressed in rectangular coordinates can be drawn. The reader must know from his/her courses in mathematics that $j = \sqrt{-1}$ and, therefore, $j \times j = -1$. The geometric interpretation of this in the complex plane is that multiplication by j rotates a phasor anticlockwise by 90°.

In Figure A.15 the phasor V of magnitude V and phase angle θ can be written in rectangular or polar form as follows: $V = a + jb = V \angle \theta$. The conversion between rectangular and polar is carried out through

$$V = \sqrt{a^2 + b^2} \quad \text{and} \quad \theta = \tan^{-1}\frac{b}{a}$$

To convert from polar to rectangular:

$$a = V \cos \theta \quad \text{and} \quad b = V \sin \theta$$

Figure A.14 Complex number plane.

Figure A.15 Rectangular–polar conversion.

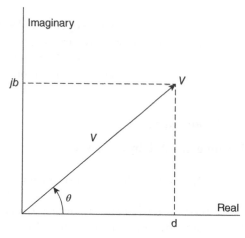

It is also possible to write $V = V(\cos\theta + j\sin\theta)$. A famous expression known as Euler's theorem states that $\cos\theta + j\sin\theta = e^{j\theta}$; therefore, an alternative expression for V in polar form is

$$V = Ve^{j\theta} \tag{A.17}$$

Circuit calculations involving the addition or subtraction of complex numbers are carried out conveniently using rectangular coordinates. Polar coordinates are more convenient when multiplications and divisions are required. Since complex numbers are legitimate mathematical entities, just like scalar numbers they can be added, subtracted, multiplied, divided, squared, inverted and so on, just like any other kind of number.

Addition and subtraction with complex numbers in rectangular form is easy. For addition, the real components are simply added to determine the real component of the sum and the imaginary components are added to determine the imaginary component of the sum. For subtraction, just take the difference of these numbers. When multiplying complex numbers in polar form, simply multiply the polar magnitudes of the complex numbers to determine the polar magnitude of the product and add the angles of the complex numbers to determine the angle of the product. If dividing, just divide the magnitudes and subtract the angles. Meter measurements in an AC circuit correspond to the *polar magnitudes* of calculated values. Rectangular expressions of complex quantities in an AC circuit have no direct, empirical equivalent, although they are convenient for calculations.

All the rules and laws of DC circuits, that is, Ohm's law, Kirchhoff's laws and network analysis methods, apply to AC circuits as well, with the exception of power calculations. The only qualification is that all variables *must* be expressed in complex form, taking into account phase as well as magnitude, and all voltages and currents must be of the same frequency (in order that their phase relationships remain constant).

A.9 Reactance and Impedance

Now expressions will be developed that describe the 'resistance' or 'opposition' offered to the flow of AC currents by the three types of circuit elements. This can be found by dividing the phasor of voltage by the resulting current phasor. The voltage will be taken as the reference phasor and polar notation will be used in these derivations.

A.9.1 Resistance

Dividing Eq. (A.4) by (A.5) gives

$$\text{Opposition offered by resistance} = \frac{V\angle 0°}{I\angle 0°} = R \ \text{ohm} \tag{A.18}$$

The relationship here is very simple. To derive the current through a resistor just divide the voltage phasor by the resistance. The current and the voltage are in phase.

A.9.2 Inductance

Dividing Eq. (A.10) by (A.11),

$$\text{Opposition offered by inductance} = \frac{V \angle 0°}{I \angle - 90°} = \frac{V \angle 0°}{V/(\omega L) \angle - 90°}$$

$$= \omega L \angle 0° - (-90)° = X_{\text{L}} \angle 90° = jX_{\text{L}} \quad \text{ohm} \tag{A.19}$$

The quantity $jX_{\text{L}} = j\omega L = j2\pi fL$ is known as *inductive reactance* and represents the 'opposition' to the flow of current when a voltage of frequency f Hz is applied across an inductance of L henry. The 90° phase lag that occurs to the current with respect to the voltage in an inductive element is conveniently incorporated in the description of the reactance through the operator j. There is nothing 'imaginary' of course about a reactor that consists typically of a coil of wire. However, as will be shown later, by assigning the j operator to the reactance the calculations are enormously assisted.

A.9.3 Capacitance

Dividing Eq. (A.14) by (A.15),

$$\text{Opposition offered by capacitance} = \frac{V \angle 0°}{I \angle 90°} = \frac{V \angle 0°}{V \omega C \angle 90°}$$

$$= \frac{1}{\omega C} \angle 0° - 90° = X_{\text{C}} \angle - 90° = -jX_{\text{L}} \quad \text{ohm} \tag{A.20}$$

The quantity $-jX_{\text{C}} = -j/(\omega C) = -j/(2\pi fC)$ is known as *capacitive reactance* and represents the 'opposition' to the flow of current when a voltage of frequency f Hz is applied across a capacitance of C farad. Again, the 90° phase lead to the current is conveniently incorporated in the $-j$ operator of the reactance expression.

A.9.4 Impedance

Now two components will be connected in series and the effects will be investigated. Take, for example, the circuit in Figure A.16.

The resistor will offer 5 Ω of resistance to AC current regardless of frequency, while the inductor will offer $2\pi fL = 6.282\,\Omega$ of reactance to AC current at 50 Hz. Because the resistor's resistance is a real number $(10 + j0\,\Omega)$ and the inductor's reactance is an imaginary number $(0 + j6.282\,\Omega)$, the combined effect of the two components will be an opposition to current equal to the complex sum of the two numbers. In order to express this opposition succinctly, a more comprehensive term is needed for opposition to the current than either resistance or reactance alone. This term is a complex number

Figure A.16 Series AC circuit.

known as *impedance Z*, and is also given in ohms, just like resistance and reactance. In the previous example, the total circuit impedance is

$$Z = (10 + j0) + (0 + j6.282) = 10 + j6.282 \text{ or } 11.81\angle 32.137°$$

Impedance is related to voltage and current in a manner similar to resistance in Ohm's law:

$$V = IZ \tag{A.21}$$

where all the quantities are expressed in complex form.

Impedance is a far more comprehensive expression of opposition to the flow of electrons than resistance. Any resistance and any reactance, separately or in combination (series/parallel), can be represented as a single impedance in an AC circuit.

To calculate current in the previous circuit, first the phase angle reference of zero is given to the voltage source. The phase angles of resistive and inductive impedance are *always* 0° and +90°, respectively, regardless of the given phase angles for voltage or current:

$$I = \frac{V}{Z} = \frac{10\angle 0°}{11.81\angle 32.137} = 0.846\angle -32.137$$

As with the purely inductive circuit, the current sine wave lags behind the voltage sine wave, although this time the angle lag is only 32.137°.

The same rules applied in the analysis of DC circuits apply to AC circuits as well, with the caveat that all quantities must be represented and calculated in complex rather than scalar form. So long as phase shift is properly represented in the calculations, there is no fundamental difference in how basic AC circuit analysis is approached versus DC.

Now a generalised expression of the complex impedance of a series circuit containing R, L and C will be written:

$$Z = R + jX = R + jX_L - jX_C = R + j(X_L - X_C) = Z\angle\theta \tag{A.22}$$

where

$$Z = \sqrt{R^2 + (X_L - X_C)^2} \text{ and } \theta = \tan^{-1}\frac{X_L - X_C}{R}$$

For an applied voltage V, the current in the circuit is given simply by $I = V/Z$. For a circuit containing a number of parallel branches Z_1, Z_2, Z_3 and so on, the equivalent impedance is given by

$$Z = \frac{1}{1/Z_1 + 1/Z_2 + 1/Z_3} \tag{A.23}$$

A.10 Power in AC Circuits

The foundations have now been set for the investigation of power flows in power system networks. In Section A.2, the ideas of generators and consumers of energy were investigated. The concepts are very clear when the quantities involved are considered at one instant in time or are of a DC nature. An investigation will be made of what happens when v_{AB} and i_{AB} are sinusoidal quantities in circuits consisting of generators, consumers plus R, L and C elements. There are partial answers to these questions for purely resistive or reactive elements. It stands to reason that in mixed circuits the power is likely be a double-frequency sinusoid partly displaced about the time axis exhibiting both an average value as well as an oscillating component. In what follows, this will be shown analytically.

Figure A.17 shows the waveforms of voltage, current, and power associated with a resistive–inductive element. As expected, the current lags the voltage by an angle more than 0° and less than 90°. The power is the instant-by-instant product of the voltage and current and has both positive and negative values. Hence, during parts of the cycle, the element acts as a consumer and during other parts as a generator of power. What decides whether the element is a consumer or generator of power over the long run is the sign of the integral:

$$P = \frac{1}{T} \int_t^{t+T} p \, dt \tag{A.24}$$

Let $v_{AB} = \widehat{V} \sin \omega t$ and $i_{AB} = \widehat{I} \sin(\omega t - \theta)$. Then

$$p = v_{AB} i_{AB} = \widehat{V} \sin \omega t \, \widehat{I} \, \sin(\omega t - \theta)$$

Using a trigonometric identity,

$$p = \frac{1}{2} \widehat{V} \, \widehat{I} \cos \phi - \frac{1}{2} \widehat{V} \, \widehat{I} \, \cos(2\omega t - \theta)$$

Using Eq. (A.2) to convert the peak into RMS quantities gives

$$p = VI \cos \phi - VI \cos(2\omega t - \theta)$$

Introducing this expression into the integral of Eq. (A.24) and integrating between the limits gives

$$P = VI \cos \theta \tag{A.25}$$

as the second term represents a symmetrical sinusoid, the average value of which over a period T is zero.

Equation (A.25) is of great significance in power systems engineering. The product VI is known as *apparent* power and is measured in *volt-amperes*. The term $\cos \theta$ is known as the *power factor* and θ as the *power factor angle*. It is the product of the power factor

Figure A.17 Power in a mixed circuit.

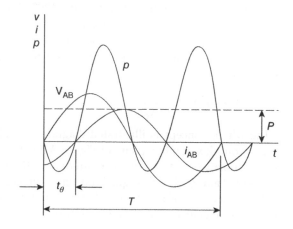

with the apparent power that results in the *active power* P that is measured in *watts*. The power factor is unity when the voltage and current are in phase, that is, for a purely resistive element (Eq. (A.7)). These are the only circumstances when the volt-amperes are equal to the watts.

The apparent power, being the product of two RMS scalar quantities that are always positive, is also always positive. It follows that the sign of Eq. (A.25), that is, whether the element is a consumer or a generator of power, depends solely on the sign of cos θ. It should be clear from Figure A.17 that as the phase-shift angle θ of the current sine wave i_{AB} increases, the average power P will decrease as the power waveform p becomes increasingly symmetrical about the time axis. When θ reaches 90°, cos $\theta = 0$ and $P = 0$. For $\theta > 90°$ the average power will reverse sign and the element will generate rather than consume power. It can then be postulated that for $-90° < \theta < 90°$ the element is a consumer of power and that for $90° < \theta < 270°$ the element is a generator of power. When $\theta = 90°$ or $-90°$, the element acts as a pure inductor or capacitor respectively and it is neither generating nor consuming active power.

A.11   Reactive Power

Equation (A.12) gives the instantaneous power for an inductor as $p = (\omega L)I^2 \sin 2\omega t$. It is known that $\omega L = X_L$ and, therefore, the peak of the power variation is given by

$$I^2 X_L = \frac{V^2}{X_L} = I^2 Z \sin \theta = I^2 \frac{V}{I} \sin \theta = Q = VI \sin \theta \tag{A.26}$$

The quantity Q is known as reactive power because of its similarity to the active power $P = I^2 R$. It is measured in *reactive volt-amperes* denoted as VAR, pronounced 'vars'. The reactive power Q is of considerable importance in power systems. Notwithstanding the fact that it has zero average value, it represents real reciprocating energy transfers between storage elements that are unavoidably parts of power system networks. These transfers are important because they result in extra energy loss in transmission lines (affecting efficiency) and in network voltage changes (affecting adversely, at times, the voltage at consumer terminals).

As with Eq. (A.25) for active power, the sign of sin θ in Eq. (A.26) indicates whether Q is positive or negative. For $0° < \theta < 180°$ the current lags the voltage, sin θ is positive and, by analogy to the active power notation, the element is said to be a consumer or a sink of reactive power. For $0° > \theta > 180°$, the current leads the voltage, sin θ is negative and the element is said to be a generator of Q. For a capacitive reactance,

$$Q = -I^2 X_C = -\frac{V^2}{X_C}$$

The whole concept is illustrated diagrammatically in Figure A.18, which is similar to the complex number plane of Figure A.14. With the current as the reference phasor, the position of V on the diagram defines the nature of a power system element in terms of P and Q. The first quadrant represents elements that are consumers of P and Q, that is, consist of resistive–inductive components, while the fourth quadrant is for resistive–capacitive elements. Quadrants 2 and 3 require generation of active power

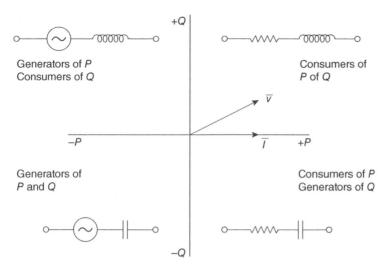

Figure A.18 Quadrant diagram.

and are reserved for 'active' elements such as AC generators. In Chapter 4, it was shown that such a device, known as a synchronous machine, is capable of operating in any of the four quadrants.

A.12 Complex Power

The *V* and *I* phasors in Figure A.18 are shown again in Figure A.19b in the form of a triangle. With the current as reference, the voltage *V* applied across the circuit is equal to the phasorial addition of the voltage V_R across the resistor (in phase with the current) and voltage V_X across the reactor (in quadrature and leading the current). If the magnitude of each voltage phasor is divided by *I*, a similar triangle, the *impedance triangle*, is obtained, shown in Figure A.19a. If each voltage magnitude is multiplied by *I*, another similar triangle, the *power triangle*, is obtained, shown in Figure A.19c. The hypotenuse of this triangle is, of course, the apparent power seen earlier. The apparent power *S* in VA can be visualised as a complex quantity with a real component $P = VI \cos \theta$ in watts and an imaginary component $Q = VI \sin \theta$ in VAR. The complex power *S* is, therefore,

Figure A.19 The complex power triangle.

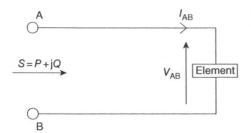

Figure A.20 Reference directions for *V*, *I*, *S*, *P* and *Q*.

defined as

$$S = VI \cos \theta + jVI \sin \theta = P + jQ \tag{A.27}$$

Note that the power factor angle θ is the same angle in all three triangles.

Now the complex power is ready to be derived directly from the applied voltage and the resulting current in a circuit element. In the element of Figure A.20, the phasors of voltage and current can be expressed in general by the exponential form

$$\boldsymbol{V}_{AB} = V_{AB}e^{j\alpha} \quad \text{and} \quad \boldsymbol{I}_{AB} = I_{AB}e^{j\beta}$$

where α and β are the angles by which $\boldsymbol{V}_{AB}$ and $\boldsymbol{I}_{AB}$ lead a reference phasor. Assume that $\alpha > \beta$. The angle by which $\boldsymbol{V}_{AB}$ leads $\boldsymbol{I}_{AB}$ (the power factor angle) is $(\alpha - \beta)$.

It would seem sensible that the multiplication of these two phasors gives the complex power. This task will now be carried out:

$$V_{AB}e^{j\alpha}I_{AB}e^{jb} = V_{AB}I_{AB}e^{j(\alpha+\beta)} = V_{AB}I_{AB}\cos(\alpha+\beta) + jV_{AB}I_{AB}\sin(\alpha+\beta)$$

Unfortunately, this product gives the wrong result! In fact, there is no good reason why an algebraic manipulation would invariably produce a result that complies with the adopted conventions. An alternative product will be tried using the *conjugate* of the current phasor $\boldsymbol{I}_{AB}^{*}$ in which the sign of the angle is negated:

$$S = V_{AB}I_{AB}^{*} = V_{AB}e^{j\alpha}I_{AB}e^{-jb} = V_{AB}I_{AB}e^{j(\alpha-\beta)}$$
$$= V_{AB}I_{AB}\cos(\alpha-\beta) + jV_{AB}I_{AB}\sin(\alpha-\beta)$$

This gives the correct result; hence, it can be confirmed that in general

$$S = VI^{*} = P + jQ \tag{A.28}$$

The accepted reference directions of voltage, current, P and Q are shown in Figure A.20, which depicts any power network element. If the complex product of $\boldsymbol{V}_{AB}$ and $\boldsymbol{I}_{AB}^{*}$ results in positive P and Q (i.e. the active and reactive power flow into the element terminals), the element is a consumer of P and Q and operates in the first quadrant of Figure A.18. In Chapters 4–6 it is shown that, depending on its nature, an element could operate in any of the four quadrants.

A.13 Conservation of Active and Reactive Power

Unlike voltage and current in AC systems, active power is a scalar quantity. If a heating element and an induction motor are connected at the terminals of a consumer, the total active power absorbed from the mains is the scalar sum of the two active powers associated with the two components. This is, of course, demanded by the conservation of energy principle.

By analogy to active power conservation, a similar notion applies to reactive power. An element consisting of an inductor and a capacitor absorbs a total reactive power equal to the sum of the component reactive powers. If the Q of the inductor is larger than the Q of the capacitor, the grid supplies reactive power to the L–C combination. If the capacitive Q is larger than the inductive Q, reactive power is supplied to the grid. In the first case the L–C combination is a net consumer of Q, in the second case it is a net generator.

These considerations are summarised by the following equations, which apply to any node, bounded portion or the totality of a power system network under steady state conditions:

$$P_{\text{entering}} - P_{\text{leaving}} - P_{\text{loss}} = 0 \Rightarrow \sum P = 0$$
$$Q_{\text{entering}} - Q_{\text{leaving}} - Q_{\text{loss}} = 0 \Rightarrow \sum Q = 0 \qquad \text{(A.29)}$$

The P_{loss} and Q_{loss} refer to any consumption of P and Q in the transmission system (see the next section).

In Chapter 5, it is shown that the concepts underlying Eq. (A.29) are vital in the calculation of power flows in power system networks.

A.14 Effects of Reactive Power Flow – Power Factor Correction

Figure A.21 represents a basic circuit of energy transportation from a generator to a consumer through a transmission line, which is simply represented by a series inductance and resistance. The consumer is represented by an inductive resistive impedance. The reason for these representations is explained in Chapter 5.

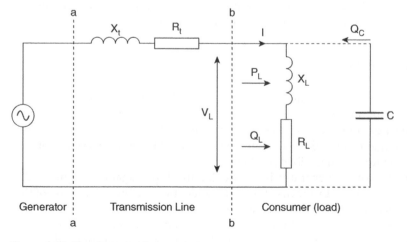

Figure A.21 Simple transmission system.

The consumer 'absorbs' both P_L and Q_L. The generator or 'the utility' has to supply both the P_L of the load and the transmission loss given by I^2R_t, where R_t is the resistance of the transmission line. It also has to supply the Q_L of the load plus the I^2X_t absorbed by the reactance of the transmission line. The transmission loss is, therefore, strongly influenced by the transfer of Q_L.

To appreciate this, assume that the load consists solely of an inductor, which absorbs Q but not P. In this case the energy meter at the consumer premises, which records kilowatt-hours (i.e. energy purchased), indicates zero but the finite current in the transmission system results in I^2R_t, which has to be supplied by the utility. Clearly, this is a most undesirable scenario that the utility endeavours to discourage, in the case of large consumers, through special tariffs that penalise the absorption of Q.

Additionally, the utility has the legal obligation to supply power to the consumer at a more or less fixed voltage. In Chapter 5, it is shown that for transmission lines, with $X_t > R_t$, the voltage V_L at the consumer terminals is particularly sensitive to changes in Q rather than P. It should now be obvious that the utility encourages consumers to draw P at minimum Q, that is, with $\cos \phi$ as close to unity as possible.

If a capacitor C were to be connected across the load terminals and sized to generate Q_C so that $Q_C = Q_L$, then the consumer will appear to the utility as having a unity power factor. The utility would be delighted with such an arrangement as this will minimise losses in the transmission line resistance R_t and will ensure minimal variation in consumer voltage. This procedure is known as *power factor correction*. Such capacitors are used extensively in power systems to generate or 'inject' reactive power at strategic points of the network.

This discussion also highlights why power system equipment is rated in terms of apparent power VA rather than P or Q. As an example, take the transmission line in Figure A.21. This line is suspended from insulators that have been designed to withstand (including a safety factor) the nominal voltage of transmission V_L. It is also made of a conductor that has a specific resistance per km and chosen so that when the rated current is carried, the heat generated does not exceed a level at which the expansion of the line produces an unacceptable sag at the lowest point of the catenary. With a purely inductive consumer load absorbing the rated current at nominal voltage, the transmission line will be fully loaded in spite of the fact that it transports no useful power. Similar considerations apply to other power system equipment.

A.15 Three-Phase AC

Power systems are almost universally three-phase AC. This gives a significant saving in the construction of all power system equipment from generators to transmission lines and consumer equipment compared to single-phase [1]. Even more importantly, three-phase very elegantly provides the rotating magnetic field required in synchronous and induction generators, as described in Chapter 4.

The three windings of the generator in Figure 4.2 are shown again in Figure A.22, feeding individual consumers by means of six independent wires. The orientation of the symbols at 120° intervals mimics the orientation of the respective voltage phasors. The voltage V_{ph} is a *phase voltage*, and the same voltage magnitude could be measured on either of the two other phases. Such a system would be perfectly capable of producing a rotating magnetic field, but the use of six wires is highly inefficient.

Figure A.22 Three-phase with six wires – three separate circuits.

In Figure A.23, the three conductors bunched together in Figure A.22 are merged into one common return conductor known as the neutral. With the circuits combined in this way, a voltage can now be measured between any two outer lines. This is known as the line-to-line voltage or simply the *line voltage* and magnitude-wise is the same no matter which two lines are chosen.

When discussing three-phase electricity it is normal practice to quote line voltages rather than phase voltages. Thus, the line voltage in Figure A.23 is indicated by V (without a subscript). Consideration of the geometry of the related phasor diagram shows that (under balanced conditions)

$$V = \sqrt{3}V_{ph}$$

where V is the line voltage and V_{ph} is the phase voltage. The four-wire arrangement is very widely used at the low voltage distribution part of a power system (Figure 1.12); for example, $V = 400\,V$ and $V_{ph} = 230\,V$.

When a three-phase system is perfectly balanced, the currents in the three phases have equal magnitudes and angles at exactly 120° intervals. Thus, the phasor sum of these currents is zero and so the current in the neutral is also zero. In this case, the neutral can be removed and the system operated with just three wires as shown in Figure A.24.

In practice, perfect balance of the three phases is not an absolute requirement. A three-wire system can operate in an unbalanced condition, but this is usually undesirable and is certainly harder to comprehend. Unbalanced operation is mentioned where necessary, but in general, balanced operation is normally assumed in this book. The

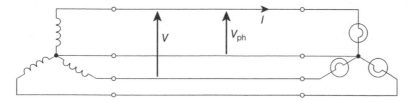

Figure A.23 Three-phase with four wires.

Figure A.24 Three-phase with three wires.

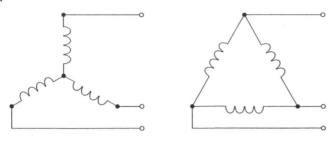

Figure A.25 Star and delta connections.

Figure A.26 Single-phase circuit used for calculations.

three-wire arrangement shown is very widely used at 'higher' voltages, that is, 11 kV and above.

The three windings of a three-phase generator can be connected either in *star* or *delta*, as shown in Figure A.25. Three-phase loads and transformers are similarly configured, and it is normal to have a mixture of star-connected and delta-connected equipment in any three-phase system.

The star configuration can provide a neutral connection for use in a four-wire system. The delta has only three external connections. There are subtle pros and cons associated with both star-connected and delta-connected configurations, particularly regarding unbalanced and faulted conditions, but these matters are beyond the terms of reference of this book.

So long as the three phases are perfectly balanced, it is perfectly legitimate and accurate to carry out calculations by considering just one of the three phases, taking care to include factors of 3 and root-3 appropriately. With this in mind, the three-phase circuit diagram of Figure A.24 can be reduced to a single phase, as shown in Figure A.26. It must be observed that the power in each phase is one-third of the total power, and, as noted before, that $V = \sqrt{3}V_{\text{ph}}$. Even in a three-wire system or where equipment is internally connected in delta, it is perfectly legitimate and accurate to use the previous approach: simply imagine a neutral conductor and reference voltages to it. Notice, also, that the current I is the same in all the preceding diagrams: there is no distinction between line current and phase current in this context. Contrast this with the currents in a delta, where $I_{\text{line}} = \sqrt{3}I_{\text{ph}}$.

A.16 The Thévenin Equivalent Circuit

The one-port circuit of Figure A.1 was initially assumed to be *passive*; that is, consisting of R, L or C components. In contrast, *active* networks include energy sources. A valuable method of representing active networks by simpler equivalent circuits is based on the Thévenin theorem. This theorem states that any one-port network consisting

Figure A.27 Thévenin equivalent circuit.

of passive elements and energy sources can be replaced by an ideal voltage source V_T (the Thévenin voltage) in series with an impedance Z_T (the Thévenin impedance). V_T is the open-circuit voltage of the two-port and Z_T is the ratio of the open-circuit voltage divided by the short-circuit current. This equivalence, which is of considerable value in power system analysis, is illustrated in Figure A.27. Proof of this theorem can be found in Reference [1] or any other text on electric network theory.

Reference

1 Smith, R.J. and Dorf, R.C. (1992). *Circuits, Devices and Systems*. Chichester: Wiley.

Index

Renewable Energy in Power Systems, Second Edition. David Infield and Leon Freris.
© 2020 John Wiley & Sons Ltd. Published 2020 by John Wiley & Sons Ltd.
Companion website: www.wiley.com/go/infield-renewable-energy